Topics in Ring Theory

CHICAGO LECTURES IN MATHEMATICS

Topics in Ring Theory

I. N. HERSTEIN

Chicago and London
THE UNIVERSITY OF CHICAGO PRESS

THE UNIVERSITY OF CHICAGO PRESS, CHICAGO 60637
THE UNIVERSITY OF CHICAGO PRESS, LTD., LONDON

Published 1969

Second Impression 1972

Printed in the United States of America

International Standard Book Number: 0–226–32802–3
Library of Congress Catalog Card Number: 69–17035

This book is dedicated to those of my mathematical colleagues, all over the world, who, in the face of great personal risk, have spoken out for the freedom and dignity of all human beings.

CONTENTS

PREFACE

These notes are based on a course I gave at the University of Chicago in the Spring Quarter of 1964. They were originally published in 1965 in the Lecture Notes of the Mathematics Department of the University of Chicago under the title of "Topics in Ring Theory." This present version is a reworking of these notes -- many of the proofs in the first chapters have been changed and some material has been added.

The subject matter divides itself naturally into three distinct parts. In the first three chapters I treat the structure of a simple associative ring and some of its subsets as Lie and Jordan rings and use these results to obtain theorems about the associative nature of the ring itself. Chapters 4, 5, 6 are devoted to the theorems of Goldie and some of their consequences for rings conditioned by some appropriate chain condition. The material of the last chapter is concerned with recent examples and counter-examples; we find there the Golod-Shafarevitch theorem and examples due to Bergman and Sasiada.

The overlap of the material here and in my Carus Monograph "Non-Commutative Rings" is not large; it consists of one of the proofs of Goldie's theorem, ultra-products and Posner's theorem and the Golod-Shafarevitch theorem.

CHAPTER 1

LIE AND JORDAN STRUCTURES IN SIMPLE RINGS

In this course we shall study certain aspects of the theory of associative rings. From time to time we shall have to refer to results proved in our set of notes "Theory of Rings" (University of Chicago) which we shall refer to as (N) or to Jacobson's book "Structure of Rings" referred to as (J).

We start things off with

LEMMA 1.1. *Let* R *be a ring and* $0 \neq \rho$ *a right ideal of* R. *Suppose that, given* $a \in \rho$, $a^n = 0$ *for a fixed integer* n; *then* R *has a non-zero nilpotent ideal.*

Proof. The argument we use is a variation of one given by Levitzki. We go by induction on n.

Let $a \neq 0$ be in ρ satisfying $a^2 = 0$; let $A = a\rho$. Suppose for the moment that $A \neq (0)$. If $x \in R$ then $(a + ax)^n = 0$ since it is in ρ, hence on expansion we get $(ax)^{n-1}a = 0$. Thus $(ax)^{n-1}A = (0)$. Let $T = \{x \in A \mid xA = (0)\}$; of course, T is an ideal of A. Moreover, as we have just seen, y in A implies that $y^{n-1} \in T$. Therefore in $\overline{A} = A/T$ every element satisfies $\overline{y}^{n-1} = 0$. By our induction hypothesis $\overline{A}$ has a nilpotent ideal $\overline{U} \neq (0)$. Let U be its inverse image in A; since $\overline{U}^k = (0)$, $U^k \subset T$ hence $U^{k+1} \subset TA = (0)$. Also, since $\overline{U} \neq 0$, $U \not\subset T$ whence $U \supset UA \neq (0)$. But then $UA = Ua\rho \neq (0)$ is a nilpotent right ideal of R.

Suppose then that $a \in \rho$, $a^2 = 0$ implies that $a\rho = (0)$. For any $x \in \rho$, since $x^n = 0$ we have $(x^{n-1})^2 = 0$ and so $x^{n-1}\rho = (0)$. Let $W = \{x \in \rho \mid x\rho = (0)\}$; W is an ideal of ρ. If $W = \rho$ then $\rho^2 = (0)$ and ρ would provide us with a nilpotent right ideal. If $W = \rho$ then in $\overline{\rho} = \rho/W$, $\overline{x}^{n-1} = 0$; our induction gives us a nilpotent ideal $\overline{V} \neq (0)$ in $\overline{\rho}$. If V is the inverse image of $\overline{V}$ in ρ then $V\rho \neq (0) \subset V$ and is nilpotent since V is. Again we have seen that R must have a non-zero nilpotent right ideal.

If R has a non-zero nilpotent right ideal it has (almost trivially) a non-zero nilpotent ideal. This proves the lemma.

Given any associative ring R we can induce on R, using its operations, two new structures, the <u>Lie structure</u> and the <u>Jordan structure</u> by defining the new products $[x, y] = xy - yx$ and $x\ y = xy + yx$ respectively. We propose to investigate the relationship between the associative structure of R and those induced Lie and Jordan structures.

We say that a subset A of R is a <u>Lie subring</u> of R if A is an additive subgroup such that for a, b in A, $ab - ba$ must also be in A. We similarly define a <u>Jordan subring</u> of R.

DEFINITION. Let A be a Lie subring of R. The additive subgroup $U \subset A$ is said to be a <u>Lie ideal</u> of A if whenever $u \in U$ and $a \in A$ then $[u, a] = ua - au$ is in U.

We similarly define the concept of a Jordan ideal of a Jordan subring of R.

Our first objective will be to determine the Lie and Jordan ideals of the ring R itself in the case when R is restricted to be a simple ring.

We begin with the Jordan ideals of R, which are a good deal easier to characterize.

LEMMA 1.2. *If* U *is a Jordan ideal of* R *then for all* $a, b \in U$ *and* $x \in R$, $(ab + ba)x - x(ab + ba) \in U$.

Proof. Since $a \in U$, for any $x \in R$, $a(xb - bx) + (xb - bx)a$ is in U. But

$$a(xb - bx) + (xb - bx)a = \{(ax - xa)b + b(ax - xa)\} + \{(ab + ba) - (ab + ba)x\}$$

The left side and the first term on the right side are in U, hence $x(ab + ba) - (ab + ba)x$ is also in U, proving the lemma.

From this we now obtain

THEOREM 1.1. *Let* R *be a ring in which* $2x = 0$ *implies* $x = 0$ *and suppose further that* R *has no non-zero nilpotent ideals. Then any non-zero Jordan ideal of* R *contains a non-zero (associative) ideal of* R.

Proof. Let $U \neq (0)$ be a Jordan ideal of R and suppose that $a, b \in U$. By Lemma 1.2, for any $x \in R$, $xc - cx \in U$ where $c = ab + ba$. However, since $c \in U$, $xc + cx \in U$. Adding we get $2xc \in U$ for all x, hence for $y \in R$, $(2xc)y + y(2xc) \in U$. Since $2yxc \in U$ we obtain $2xcy \in U$, that is $2RcR \subseteq U$. Now 2RcR is an ideal of R so we are done unless $2RcR = (0)$. If $2RcR = (0)$, by our assumptions $RcR = (0)$ and so $(Rc)^2 = (0)$. Since R has no nilpotent ideals this forces $c = 0$, that is, given $a, b \in U$ then $ab + ba = 0$.

Let $0 \neq a \in U$; then for $x \in R$, $b = ax + xa \in U$ hence $a(ax + xa) + (ax + xa)a = 0$. That is, $a^2x + xa^2 + 2axa = 0$. Now for $a \in U$, $0 = aa + aa = 2a^2$ whence $a^2 = 0$. The top relation then reduces to $2axa = 0$ for all $x \in R$ and so $aRa = (0)$. But then $aR \neq (0)$ is a nilpotent right ideal of R, contrary to assumption. In other words, we have shown that U contains a non-zero ideal of R.

COROLLARY. *If* R *is a simple ring of characteristic* $\neq 2$ *then* R *is simple as a Jordan ring.*

We now turn to the case of the Lie ideals of R.

DEFINITION. If A, B are subsets of R then $[A, B]$ is the additive subgroup of R generated by all $ab - ba$ with $a \in A$, $b \in B$.

LEMMA 1.3. *Let* R *be a ring with no non-zero nilpotent ideals in which* $2x = 0$ *implies* $x = 0$. *Suppose that* $U \neq (0)$ *is both a Lie ideal and a subring of* R. *Then either* $U \subset Z$, *the center of* R, *or* U *contains a non-zero ideal of* R.

Proof. Let us first suppose that U, as a ring, is not commutative. Then for some $x, y \in U$, $xy - yx \neq 0$. For any $r \in R$, $x(yr) - (yr)x$ is in U, that is $(xy - yx)r + y(xr - rx)$ is in U. The second member of this is in U since both y and $xr - rx$ are in U (since U is both a Lie ideal and subring). The net result of all this is that $(xy - yx)R \subset U$. But then for $r, s \in R$, $((xy - yx)r)s - s((xy - yx)r) \in U$ leading to $R(xy - yx)R \subset U$. We have now shown that the ideal $R(xy - yx)R$ is in U. If $R(xy - yx)R = (0)$ then $(R(xy - yx))^2 = (0)$ contrary to assumption. We have shown that the result is correct if U as a subring of R is not commutative.

So, suppose that U is commutative; we want to show that it lies in the center of R. Given $a \in U$, $x \in R$ then $ax - xa \in U$ so commutes with a. Now, for $x, y \in R$, $a(a(xy) - (xy)a) = (a(xy) - (xy)a)a$. Expanding $a(xy) - (xy)a$ as $(ax - xa)y + x(ay - ya)$ and using that a commutes with this, with $ax - xa$ and with $ay - ya$ yields $2(ax - xa)(ay - ya) = 0$ for all $x, y \in R$. Since $2r = 0$ forces $r = 0$ we obtain $(ax - xa)(ay - ya) = 0$. In this put $y = ax$; this results in $(ax - xa)R(ax - xa) = (0)$. Since R has no nilpotent ideals we conclude that $ax - xa = 0$ and so, a must be in the center of R.

Note that in the latter part of the proof of Lemma 1.3 we have also proved the

Sublemma. Let R be a ring having no non-zero nilpotent ideals in which $2x = 0$ implies that $x = 0$. If $a \in R$ commutes with all $ax - xa$, $x \in R$ then a is in the center of R.

Lemma 1.3 immediately implies

THEOREM 1.2. *Let* R *be a simple ring of characteristic* $\neq 2$. *Then any Lie ideal of* R *which is also a subring of* R *must either be* R *itself or is contained in the center of* R.

DEFINITION. If U is a Lie ideal of R let $T(U) = \{x \in R \mid [x, R] \subset U\}$.

LEMMA 1.4. *For any ring* R, *if* U *is a Lie ideal of* R, *then* $T(U)$ *is both a subring and a Lie ideal of* R; *moreover,* $U \subset T(U)$.

Proof. Since U is a Lie ideal of R, $U \subset T(U)$; since $[T(U), R] \subset U \subset T(U)$, $T(U)$ must certainly be a Lie ideal of R.
Now suppose that $a, b \in T(U)$, $r \in R$. Then

$$(ab)r - r(ab) = \{a(br) - (br)a\} + \{b(ra) - (ra)b\},$$

so, since $a, b \in T(U)$, the right side is in U. Therefore $[ab, R] \subset U$, that is $ab \in T(U)$.

We now prove the

THEOREM 1.3. *Let* R *be a simple ring of characteristic* $\neq 2$ *and let* U *be a Lie ideal of* R. *Then either* $U \subset Z$, *the center of* R, *or* $U \supset [R, R]$.

Proof. By Theorem 1.2 and Lemma 1.4, since $T(U)$ is both a subring and a Lie ideal of R, either $T(U) \subset Z$ or $T(U) = R$. If $T(U) = R$ then by its very definition $[R, R] \subset U$; if $T(U) \subset Z$, since $U \subset T(U)$, we obtain $U \subset Z$.

COROLLARY. *If* R *is a non-commutative simple ring of characteristic* $\neq 2$, *then the subring generated by* $[R, R]$ *is* R.

Proof. Any additive subgroup containing $[R, R]$ is, trivially, a Lie ideal of R. Hence the subring generated by $[R, R]$ is a Lie ideal thus, by Theorem 1.2, it equals R or is in Z. If it is in Z then $[R, R] \subset Z$. Thus for $a \in R$, a commutes with all $ax - xa$, $x \in R$; by the sublemma we get that $a \in Z$, that is, $R \subset Z$. Since we assumed R to be non-commutative, this is ruled out; hence the corollary.

We now should like to settle the problem even when R has characteristic 2. Note that the characteristic of R has not entered into the discussion in the passage from Theorem 1.2 on. So we ask: when, in characteristic 2, does Theorem 1.2 fail? It certainly fails in F_2, the ring of all 2 by 2 matrices over F, a field of characteristic 2 for $U = \{\begin{pmatrix} a & b \\ b & a \end{pmatrix} \mid a, b \in F\}$ is a Lie ideal and subring of R which is neither in Z nor does it equal R. We aim to show that this is, effectively, the only counter-example.

Suppose that R is a simple ring of characteristic 2 and that U is a Lie ideal and subring of R, $U \neq R$ and $U \not\subset Z$. As in the proof of Lemma 1.3 we obtain that U, as a subring of R, must be commutative. That is, given $u, v \in U$ then $uv + vu = 0$.

Let $a \in U$; then $as + sa \in U$ for all $s \in R$ hence $a(as + sa) = (as + sa)a$. This says that $a^2 \in Z$. Since for any $r \in R$, $ar + ra \in U$ we also have that $(ar + ra)^2 \in Z$.

If $Z = (0)$ then $a^2 = 0$, $(ar + ra)^2 = 0$ from which we get $(ar)^3 = 0$. But then aR is a right ideal of R in which the cube of every element is 0; by Lemma 1.1 R would have a nilpotent ideal, that is, R would be nilpotent, which is impossible for a simple ring.

Therefore we may assume that $Z \neq (0)$ and that there is an element $a \in U$, $a \notin Z$ such that $a^2 \neq 0 \in Z$ and $(ar + ra)^2 \in Z$ for all $r \in R$.

To answer completely what the structure of R must be we prove a subsidiary

THEOREM 1.4. *Let* R *be a simple ring of characteristic* 2 *and suppose that there exists an* $a \in R$, $a \notin Z$ *such that* $a^2 \in Z$ *and* $(ax + xa)^4 \in Z$ *for all* $x \in R$. *Then* R *is* 4-*dimensional over* Z.

Before proving the theorem we should like to point out that a more general theorem actually holds, namely: if R is a simple ring with an element $a \notin Z$ such that $(ax - xa)^n \in Z$ for all $x \in R$ then R is 4-dimensional over Z.

Proof. of Theorem 1.4. If $Z = (0)$ then both $a^2 = 0$ and $(ax + xa)^4 = 0$ hence $(ax)^5 = a(ax + xa)^4 x = 0$ for all $x \in R$. But then the right ideal aR satisfies $u^5 = 0$ for all elements $u \in aR$; by Lemma 1.1 this is not possible in a simple ring.

Suppose, then, that $Z \neq (0)$, hence $1 \in R$. If $a^2 = 0$ then $b = a + 1$ satisfies $b^2 = 1$ and $(bx + xb)^4 \in Z$ for all $x \in R$. Therefore we may assume that $a^2 = \alpha \neq 0$ in Z. Let $Z' = Z(\sqrt{\alpha})$; then $R' = R \otimes_Z Z'$ is simple. Moreover, in R' we have $(ax' + x'a)^4 \in Z'$ for all $x' \in R'$. Since $\dim R'/Z' = \dim R/Z$, to prove the theorem it is enough to do so in R'. Also $b = a/\beta$ where $\beta \in Z'$, $\beta^2 = \alpha$ satisfies $b^2 = 1$ and $(bx' + x'b)^4 \in Z$. Hence, without loss of generality we may suppose that $a \in R$, $a \notin Z$, $a^2 = 1$ and $(ax + xa)^4 \in Z$ for all $x \in R$.

Now R is a dense ring of linear transformations on a vector space V over a division ring D (since $Z \neq (0)$ and R is simple). Since $(a + 1)^2 = 0$, $a + 1 \neq 0$, V must be more than 1-dimensional over D. Since $a \neq 1$ it is immediate that there is a $v \in V$ such that v, va are linearly independent over D.

If for some $w \in V$, v, va and $w(1+a)$ are linearly independent over D then the subspace V_o spanned by these is invariant under a and a induces the linear transformations

$$\begin{pmatrix} 0 & 1 & 0 \\ 1 & 0 & 0 \\ 0 & 0 & 1 \end{pmatrix}$$

on V_o. By density of R on V there is an $x \in R$ which induces

$$\begin{pmatrix} 0 & 1 & 0 \\ 0 & 0 & 0 \\ 0 & 0 & 0 \end{pmatrix}$$

on V_o hence $ax + xa$ induces

$$\begin{pmatrix} 0 & 1 & 0 \\ 1 & 0 & 0 \\ 0 & 0 & 0 \end{pmatrix}$$

on V_o. But $(ax + xa)^4 \in Z$ yet does not induce a scalar on V_o since it induces

$$\begin{pmatrix} 1 & 0 & 0 \\ 0 & 1 & 0 \\ 0 & 0 & 0 \end{pmatrix}$$

Thus we have that for all $w \in V$, v, va and $w(1+a)$ are linearly dependent over D.

If V is more than 2-dimensional over D there is a $w \in V$ such that v, va, w are linearly independent over D. By the above, wa is in the subspace V_1 they span. The matrix of a on V_1 is

$$\begin{pmatrix} 0 & 1 & 0 \\ 1 & 0 & 0 \\ \alpha & \beta & \gamma \end{pmatrix}$$

By density there is an $x \in R$ which induces $\begin{pmatrix} 0 & 1 & 0 \\ 0 & 0 & 0 \\ 0 & 0 & 0 \end{pmatrix}$ on V_1;

but then $ax + ax$ induces

$$\begin{pmatrix} 1 & 0 & 0 \\ 0 & 1 & 0 \\ 0 & \alpha & 0 \end{pmatrix}$$

whose fourth power is not a scalar.

Thus we must have that V is 2-dimensional over D. All that remains is to show that D is commutative.

Let $a = \begin{pmatrix} \alpha & \beta \\ \gamma & \delta \end{pmatrix}$; then $a^2 = 1$ yields $\alpha^2 + \beta\gamma = 1$, $\alpha\beta + \beta\delta = 0$, $\gamma\beta + \delta^2 = 1$. In particular not both $\alpha = 0$ and $\gamma = 0$. If $\sigma \in D$ then using $x = \begin{pmatrix} 0 & \sigma \\ 0 & 0 \end{pmatrix}$ and $(ax + xa)^4 \in Z$ yields

$$\begin{pmatrix} \sigma\alpha & \alpha\sigma + \sigma\beta \\ 0 & \gamma\sigma \end{pmatrix}^4 \in Z$$

This gives for all $\sigma \in D$, $(\sigma\alpha)^4 \in Z$ and $(\gamma\sigma)^4 \in Z$. If $\alpha \neq 0$ then $\sigma\alpha$ runs through D as σ does, so every $x \in D$ would satisfy $x^4 \in Z$. But a non-commutative division ring can not be purely inseparable over its center (see (N), Lemma 2.16). Thus $\alpha \neq 0$ implies D is commutative. Similarly $\gamma \neq 0$ implies that D is commutative. Since one of these must hold we get that D is commutative and so R is 4-dimensional over Z.

Since the hypothesis of Theorem 1.4 is precisely the one lead to by the assumption that Theorem 1.2 (and so Theorem 1.3) was false we obtain

THEOREM 1.5. *If* R *is a simple ring and if* U *is a Lie ideal of* R *then either* $U \subset Z$ *or* $U \supset [R, R]$ *except if* R *is of characteristic* 2 *and is* 4-*dimensional over its center.*

The theorem has as an immediate corollary the

COROLLARY. *If* R *is a simple non-commutative ring then the subring generated by* $[R, R]$ *is* R.

As a consequence of these results we prove a theorem about simple rings with involution. Let R be a simple ring of characteristic not 2 and suppose that on R there is defined an involution $*$, that is, a map of R into itself satisfying $a^{**} = a$, $(a+b)^* = a^* + b^*$, $(ab)^* = b^*a^*$. Let $S = \{a \in R \mid a^* = a\}$ and $K = \{a \in R \mid a^* = -a\}$. S is the set of <u>symmetric elements</u>, K the set of <u>skew-symmetric</u> elements of R. Note that, since the characteristic of R is not 2, $R = S + K$, $S \cap K = (0)$, and that S is a Jordan subring of R and K a Lie subring of R. Also, if $s \in S$ and $k \in K$ then $sk - ks \in S$. We shall make an intensive study of S and K in the next chapter. But now we content ourselves with proving the

THEOREM 1.6. *Let* R *be a simple ring with involution of characteristic not* 2, *and let* S *be the set of symmetric elements of* R. *Then* $\overline{S}$, *the subring of* R *generated by* S, *is* R *unless* R *is of dimension* 4 *or less (thus* 4 *or* 1) *over its center.*

Proof. We claim that $\overline{S}$ is a Lie ideal of R. To see this note first that trivially $[\overline{S}, S] \subset \overline{S}$. If $k \in K$ and $\overline{s} \in \overline{S}$ we want to show that $[\overline{s}, k] \in \overline{S}$; to do so, since $\overline{s}$ is a sum of monomials from S, we need merely do it for monomials $\overline{s} = s_1 s_2 \cdots s_n$, $s_i \in S$. But then $[s_1 s_2 \cdots s_n, k] = [s_1, k] s_2 \cdots s_n + \cdots + s_1 \cdots s_{i-1}[s_i, k] s_{i+1} \cdots s_n + \cdots + s_1 s_2 \cdots s_{n-1}[s_n, k]$ which is certainly in $\overline{S}$. Thus $[\overline{S}, R] = [\overline{S}, S + K] = [\overline{S}, S] + [\overline{S}, K] \subset \overline{S}$ and so $\overline{S}$ is a Lie ideal of R. By definition it is a subring of R. Therefore by Theorem 1.2 we conclude that either $\overline{S} = R$ or $\overline{S} \subset Z$.

We consider the second possibility, namely, $\overline{S} \subset Z$. But then $S \subset Z$. Given $a \in R$, $a = s + k$, $s \in S \subset Z$, $k \in K$ hence $(a - s)^2 = k^2 \in S \subset Z$. Consequently $a^2 - 2sa + s^2 - k^2 = 0$, which is to say, R is quadratic over Z. By Theorem 2.38 of (N) we get that R is at most 4-dimensional over Z.

Related to this theorem is the following remark which holds for simple rings, of any characteristic, which have involutions.

THEOREM 1.7. *Let R be a simple ring with involution whose center $Z = (0)$ or for which $\dim R/Z > 4$. Then the only elements commuting with all the symmetric elements lie in Z.*

Proof. Let $a \in R$ commute with all $s \in S$. If the characteristic of R is not 2, by Theorem 1.6 $\overline{S} = R$ hence $a \in Z$ follows. Thus we may suppose that R is of characteristic 2.

Let $T = \{r \in R \mid rs = sr, s \in S\}$. T is clearly a subring of R. Given $x \in R$, $a \in T$ then $a(x+x^*) = (x+x^*)a$ hence $ax + xa = ax^* + x^*a$. We want to show that T is a Lie ideal of R. Given $a \in T$, $y \in R$, $s \in S$ then $(ay + ya)s = ays + yas$ $= ays + ysa = a(ys + sy^*) + asy^* + ysa = (ys + sy^*)a + say^* + ysa$ $= sy^*a + say^* = s(y^*a + ay^*) = s(ya + ay)$, as we have just shown. In other words, T is both a Lie ideal and subring of R. By our assumption on $\dim R/Z$ we get from Theorem 1.5 and its corollary that $T \subset Z$ or $T = R$. If $T = R$ then $S \subset Z$ which we have seen forces $\dim R/Z \leq 4$. Thus $T \subset Z$, which is the assertion of the theorem.

We now turn to a study of the Lie structure of $[R, R]$ as a Lie subring of R.

We begin with an easy result,

LEMMA 1.5. *Let R be a ring without any non-zero nilpotent ideals. Then any element of R which commutes with all elements of $[R, R]$ must lie in the center of R.*

Proof. Let a commute with all elements of $[R, R]$. Given $x, y \in R$ then a commutes with $xy - yx$ and with $x(xy) - (xy)x = x(xy - yx)$. The upshot of this is that $(ax - xa)(xy - yx) = 0$ for all $x, y \in R$. In this, replace y by ya; since $x(ya) - (ya)x = (xy - yx)a + y(xa - ax)$ we get $0 = (xa - ax)(x(ya) - (ya)x) = (xa - ax)(xy - yx)a + (xa - ax)y(xa - ax)$. This results in $(xa - ax)R(xa - ax) = (0)$; but then $(xa - ax)R$ is a nilpotent right ideal so is (0). Since R has no nilpotent ideals we get from this $xa - ax = 0$, that is, a must be in the center of R.

THEOREM 1.8. *If R is a simple ring and is not 4-dimensional over a field of characteristic* 2, *then* $[[R, R], [R, R]] = [R, R]$.

Proof. If R is commutative the result is clearly correct; so suppose R is not commutative. Now, $[[R, R], [R, R]]$ is a Lie ideal of R hence, by Theorem 1.6, either $[[R,R], [R,R]] \supset [R, R]$ (the desired conclusion) or $[[R, R], [R, R]] \subset Z$. Suppose, then, that $[[R, R], [R, R]] \subset Z$.

If $[R, R] \subset Z$, then by the corollary to Theorem 1.5, R is commutative. So $[R, R] \not\subset Z$. Let $a \in [R, R]$, $a \notin Z$ and let $x \in R$. Thus $\alpha = [a, [a, x]]$ and $\beta = [a, [a, ax]]$ are in $[[R, R], [R, R]]$, hence in Z. Since $\beta = \alpha a$ and Z is a field, if $\alpha \neq 0$ then it follows that $a \in Z$. So α must be 0. Thus $[a, [a, x]] = 0$ for all $x \in R$.

If the characteristic of R is not 2 this forces $a \in Z$ on invoking the sublemma. If the characteristic of R is 2 then $[a, [a, x]] = 0$ implies $a^2 \in Z$; hence $a^2 \in Z$ for all $a \in [R, R]$. But for $x \in R$, $ax - xa \in [R, R]$ so $(ax - xa)^2 \in Z$ and $a^2 \in Z$ and $a \notin Z$. By Theorem 1.4 we get that $\dim R/Z = 4$ contrary to hypothesis. The theorem is now proved.

Let U be a Lie ideal of $[R, R]$ and set $T(U) = \{x \in R \mid [x, R] \subset U\}$. It is immediate that $T(U)$ is a subring of R. We investigate the interrelationship of $T(U)$ and U in

LEMMA 1.6. *If* U *is a Lie ideal of* $[R, R]$ *then:*

1. $[U, U] \subset T(U)$
2. $[U, T(U)] \subset T(U)$
3. $[[R, T(U)], T(U)] \subset T(U)$
4. $[[T(U), T(U)], R] \subset T(U)$.

Proof. The proofs are all of the same spirit, making use of the Jacobi identity.

Now $[[U,U],R] \subset [[U, R],U] + [U,[U,R]] \subset [U, [R, R]] \subset U$ hence $[U,U] \subset T(U)$. Also $[[U,T(U)],R] \subset [[U, R],T(U)] + [U,[R,T(U)]] \subset U$ by definition of $T(U)$, so $[U,T(U)] \subset T(U)$. Now $[R,T(U)] \subset U$ so (3) follows from (2). Finally

$[[T(U), T(U)], R] \subset [T(U), [T(U), R]] \subset [T(U), U] \subset T(U)$

by (2). This proves the lemma.

Let $A = [T(U), T(U)]$; we prove

LEMMA 1.7. $R[A, A]R \subset U + T(U)$.

Proof. Let $a, b \in A$; then for $r \in R$, $[a,b]r = [a,br] - b[a,r]$. The first term is in $T(U)$ by part 4 of Lemma 1.6, as is $[a, r]$. Since $T(U)$ is a subring of R, $b[a, r] \in T(U)$, hence $[A,A]R \subset T(U)$. Similarly $R[A,A] \subset T(U)$. Now $R[A,A]R \subset [T(U),R] + [A,A]R$ $\subset U + T(U)$.

THEOREM 1.9. *Let* R *be a simple ring and let* $U \neq [R, R]$ *be a Lie ideal of* $[R, R]$. *Suppose* R *is not* 4-*dimensional over a field of characteristic* 2. *Then*

$$[[[U, U], [U, U]], [[U, U], [U, U]]] = (0).$$

Proof. Since $U \neq [R, R]$, $T(U) \not\supset [R, R]$ for otherwise $U \supset [T(U), R] \supset [[R, R], R] = [R, R]$ by Theorem 1.8. Now

$$[U + T(U), U + T(U)] \subset [U, U] + [U, T(U)] + [T(U), T(U)] \subset T(U)$$

by Lemma 1.6. Since $T(U) \not\supset [R, R]$ we must have that $U + T(U) \neq R$. However by the previous lemma $R[A,A]R \subset U + T(U)$ and is an ideal of R; since $U + T(U) \neq R$ we get that $R[A,A]R = (0)$, hence $[A, A] = (0)$. Since $[U, U] \subset T(U)$ and $A = [T(U), T(U)]$ we get the theorem.

LEMMA 1.8. *If R is a simple ring, $\lambda \neq (0)$ a left ideal of R then $\lambda + [R, R] = R$.*

Proof. Clearly $\lambda + [R,R] \supset \lambda + [\lambda, R]$; if $a \in \lambda$, $r \in R$ then $ar = (ar - ra) + ra \in \lambda + [\lambda, R]$, that is $R = \lambda R \subset \lambda + [\lambda, R]$.

This gives us

LEMMA 1.9. *If $U \neq (0)$ is a Lie ideal of $[R, R]$, R simple, then $xU = (0)$ implies $x = 0$.*

Proof. Let $\lambda = \{x \in R \mid xU = (0)\}$; clearly λ is a left ideal of R. If $\lambda \neq (0)$ then by the previous lemma $R = \lambda + [R, R]$. Let $x \in \lambda$, $a \in [R, R]$, $u \in U$; then $(xa - ax)u = xau = x(au - ua) = 0$, that is, $[\lambda, [R, R]] \subset \lambda$. Since λ is a left ideal of R this implies that $\lambda[R, R] \subset \lambda$, hence $R = \lambda R = \lambda(\lambda + [R, R]) \subset \lambda$ forcing $RU = (0)$. This contradicts that $U \neq (0)$, so we must conclude that $\lambda = (0)$.

We now prove the basic

THEOREM 1.10. *Let R be a simple ring and U a Lie ideal of $[R, R]$ such that $[U, U] = (0)$. Then:*

1. *if the characteristic of R is not 2, $U \subset Z$.*
2. *if the characteristic of R is 2 then for every $u \in U$, $u^2 \in Z$.*

<u>Proof.</u> Let $u \neq 0 \in U$ and let $d(x)$ be defined by $d(x) = ux - xu$ for $x \in R$.

By assumption, if $x, y \in R$ then since $[u,[x,y]] \in U$, $d^2([x,y]) = 0$. In this last relation replace y by zv with $v \in U$. Using Leibnitz's rule we get

$$\begin{aligned} 0 &= d^2([x,zv]) = d^2([x,z]v + z[x,v]) \\ &= d^2(z[x,v]) \quad (\text{since } v \in U,\ d(v) = 0 \text{ and } d^2([x,z]) = 0 \text{ so } d^2([x,z]v) = 0\,) \\ &= d^2(z)[x,v] + 2d(z)d([x,v]) + zd^2([x,v]) \\ &= d^2(z)[x,v] + 2d(z)d([x,v]). \end{aligned}$$

If $x \in [R,R]$ then $[x,v] \in U$ hence $d([x,v]) = 0$, thus $d^2(z)[U,[R,R]] = (0)$ for all $z \in R$. Since $[U,[R,R]]$ is a Lie ideal of $[R,R]$, were it not (0), we would get by Lemma 1.9 that $d^2(z) = 0$.

On the other hand if $[U,[R,R]] = (0)$ by Lemma 1.5 U would be in Z. So we may suppose that $d^2(z) = 0$ for all $z \in R$, that is, $u \in U$ commutes with all $ux - xu$. By the sublemma following Lemma 1.3 we get that if the characteristic of R is not 2 then $U \subset Z$. If the characteristic is 2 then from $u(ux+xu) = (ux+xu)u$ we get $u^2x = xu^2$ for all $x \in R$ and so $u^2 \in Z$. This proves the theorem.

We sharpen the last theorem to

THEOREM 1.11. <u>If</u> R <u>is simple of characteristic not</u> 2 <u>and if</u> U <u>is a Lie ideal of</u> $[R,R]$ <u>such that</u> $[U,U] \subset Z$ <u>then</u> $U \subset Z$.

<u>Proof.</u> If $[U,U] = (0)$ then by Theorem 1.10 we are done. So suppose that there are $u, v \in U$ such that $\alpha = [u,v] \neq 0$; by assumption $\alpha \in Z$. Let $d(x) = ux - xu$ for $x \in R$. By assumption, $d(v) = \alpha \neq 0 \in Z$.

If $d^2(x) = 0$ for all $x \in R$ then $d^2(xv) = 0$; expanding, using $d^2(v) = 0$, we get $0 = d^2(x)v + 2\alpha\, d(x) = 2\alpha\, d(x)$. Since $2\alpha \neq 0 \in Z$ it is invertible so $d(x) = 0$ follows, for all $x \in R$. This contradicts that $d(v) = \alpha \neq 0$. We claim that $d^2(x) \in Z$ for all $x \in R$ is also impossible. For suppose $d^2(y) = \beta \neq 0 \in Z$; then $d^2(uy) = ud^2(y) = u\beta \in Z$ leading to $u \in Z$.

So we can pick $x \in R$ so that $d^2(x) \notin Z$. Since U is a Lie ideal of $[R,R]$, $d^2(x) \in U$ hence $[u, d^2(x)] = \gamma \in [U,U] \subset Z$. By the same token $[v, d^2(ux)] = \delta \in Z$; but $\delta = [v, d^2(ux)] = [v, ud^2(x)]$ $= [v,u]\, d^2(x) + u[v, d^2(x)] = -\alpha d^2(x) + \delta u$. Since $\alpha \neq 0$ we get $d^2(x) = \mu + \lambda u$; because $d^2(x) \notin Z$, $\lambda \neq 0$. Similarly $d^2(ux) = \mu_1 + \lambda_1 u$, $\mu_1, \lambda_1 \in Z$. But $\mu_1 + \lambda_1 u = d^2(ux) = ud^2(x) = u(\mu + \lambda u) = \lambda u^2 + \mu u$. Since $\lambda \neq 0$ we get that $u^2 + \sigma u + \tau = 0$ for some $\sigma, \tau \in Z$.

Thus $0 = [u^2 + \sigma u + \tau, v] = [u^2, v] + \sigma[u,v] = u(uv - vu) + (uv - vu)u + \alpha\sigma$ $= 2\alpha u + \alpha\sigma$. Since $2\alpha \neq 0$ we get from this the contradiction $u \in Z$. Thus the theorem is proved, and $U \subset Z$.

We are now in a position to prove one of the most fundamental theorems in this area of results, namely

THEOREM 1.12. Let R be a simple ring of characteristic not 2; then any Lie ideal of $[R,R]$ is either $[R,R]$ or is contained in the center of R.

Proof. Let $U \neq [R,R]$ be a Lie ideal of $[R,R]$. According to Theorem 1.9, $[[[U,U],[U,U]],[U,U],[U,U]] = (0)$, hence by Theorem 1.10, $[[U,U],[U,U]] \subset Z$. By Theorem 1.11 $[U,U] \subset Z$ and so again by use of the same theorem $U \subset Z$.

We now consider the case of characteristic 2. Suppose that U is a Lie ideal of $[R,R]$ such that $[U,U] \subset Z$. Given $u \in U$, $x \in [R,R]$ then $u(ux+xu) + (ux+xu)u \in Z$, that is $u^2x + xu^2 \in Z$. Let $W = \{a \in R \mid [a,[R,R]] \subset Z\}$; W is a Lie ideal of R so if

$\dim R/Z > 4$ by Theorem 1.5 $W \subset Z$ or $W \supset [R,R]$. However, if $W \supset [R,R]$ then $[[R,R],[R,R]] \subset Z$, so by Theorem 1.8, $[R,R] \subset Z$, thus since $U \subset [R,R]$ we have $U \subset Z$. If $W \subset Z$ then $a \in U$ implies $a^2 \in Z$ since $a^2 \in W$.

If $a \in U$ and $x \in R$ then $a(x(ax)+(ax)x) + (x(ax)+(ax)x)a \in U$, that is, $(ax)^2 + a^2x^2 + (xa)^2 + ax^2a \in U$. But since $a^2 \in Z$, $(ax)^2 + a^2x^2 + (xa)^2 + ax^2a = (ax+xa)^2$; since $(ax+xa)^2 \in U$ then $(ax+xa)^4 \in Z$ for all $x \in R$. By Theorem 1.4 if $\dim R/Z > 4$ this forces $a \in Z$, that is $U \subset Z$. We have sharpened Theorem 1.12 to its definitive form.

THEOREM 1.13. *If* R *is a simple ring and if* U *is a Lie ideal of* $[R,R]$ *then* $U = [R,R]$ *or* $U \subset Z$ *except if* R *is of characteristic* 2 *and* 4-*dimensional over* Z.

If T is an additive subgroup of R such that $[T,[R,R]] \subset T$, even if $T \not\subset [R,R]$ we can still pin it down tightly as we see in

THEOREM 1.14. *If* R *is a simple ring and if* $[T,[R,R]] \subset T$ *then either* $T \subset Z$ *or* $T \supset [R,R]$ *except if* R *is of characteristic* 2 *and* 4-*dimensional over* Z.

Proof. Let $S = [T,[R,R]]$; from the assumption on T, S is a Lie ideal of $[R,R]$ therefore by Theorem 1.13 either $S \subset Z$ or $S \supset [R,R]$. In the second possibility, since $T \supset S$, we get that $T \supset [R,R]$.

If $[T,[R,R]] = S \subset Z$ let $W = \{x \in R \mid [x,[R,R]] \subset Z\}$. As is easily verified, W is a Lie ideal of R hence $W \subset Z$ or $W \supset [R,R]$. Since, as we have seen several times, $[R,R] \not\subset Z$ and since $[[R,R],[R,R]] = [R,R]$, we obtain that $W \not\supset [R,R]$. Thus we are forced to conclude that $W \subset Z$. Since $T \subset W$ we get the desired conclusion $T \subset Z$.

We now show a general construction of a Lie ideal in any ring. We shall use this to obtain a theorem of Amitsur about subspaces of a simple ring invariant with respect to a suitable set of automorphisms.

LEMMA 1.10. *In any ring* R *the additive subgroup*, E, *of* R *generated by the idempotents of* R *is a Lie ideal of* R.

Proof. Let e be any idempotent in R and x any element of R. Then, as is quickly checked, $f_1 = e + xe - exe$ and $f_2 = e + ex - exe$ are idempotents. Hence $xe - ex = f_1 - f_2$ is in the additive group E generated by the idempotents. This proves the lemma.

COROLLARY. *Let* R *be a simple ring having an idempotent* $e \neq 0,1$. *Then* $E \supset [R, R]$.

Proof. Since the center, Z, of R is (0) or a field, $e \notin Z$, hence $E \not\subset Z$. But by Lemma 1.10 E is a Lie ideal of R, hence $E \supset [R, R]$ unless R is 4-dimensional over Z and Z is of characteristic 2 (Theorem 1.13). If $\dim R/Z = 4$ and Z is of characteristic 2, since R is simple and has a non-trivial idempotent, by Wedderburn's theorem $R \approx Z_2$. But then [R, R] consists of the matrices of trace 0, that is, of matrices of the form $\begin{pmatrix} \alpha & \beta \\ \gamma & \alpha \end{pmatrix}$, $\alpha, \beta, \gamma \in Z$. But $\begin{pmatrix} \alpha & \beta \\ \gamma & \alpha \end{pmatrix} = e_1 + e_2 + e_3 + e_4$ where the e_i are the following idempotents:

$$e_1 = \begin{pmatrix} \alpha & \alpha \\ 1+\alpha & 1+\alpha \end{pmatrix}, \quad e_2 = \begin{pmatrix} 1 & \alpha+\beta \\ 0 & 0 \end{pmatrix}, \quad e_3 = \begin{pmatrix} 1 & 0 \\ 0 & 0 \end{pmatrix},$$

$$e_4 = \begin{pmatrix} 0 & 0 \\ 1+\alpha+\gamma & 1 \end{pmatrix}.$$

This establishes the corollary.

COROLLARY. *Let* R *be a simple ring having an idempotent* $e \neq 0,1$. *Then* $[E, R] \supset [R, R]$

Proof. If R is not 4-dimensional over a field of characteristic 2, since $E \supset [R, R]$, $[E, R] \supset [[R, R], R]$ and $[[R, R], R] = [R, R]$ by Theorem 1.8. If R is 4-dimensional over a field of characteristic 2, since $R = Z_2$ and, by a direct computation $[[R,R],R] = [R, R]$, the fact that $E \supset [R, R]$ implies $[E, R] \supset [R, R]$.

Given $a \in R$, $a^2 = 0$ the mapping

$$x \to (1+a)x(1-a) = x + ax - xa - axa$$

is an automorphism of R. We call such an automorphism a *special inner automorphism* of R.

If R is a simple ring it is an algebra over its centroid, which is a field; if, in addition, R has a unit element its centroid coincides with Z (see (N), p. 53-54). By a subspace of R we shall mean a subset of R which is a vector space over the centroid of R.

We now prove

THEOREM 1.15. *Let* R *be a simple ring whose centroid* F *is not* GF(2); *suppose further that* R *has an idempotent* $e \neq 0,1$. *Then any subspace of* R *invariant under all special inner automorphisms of* R *must be* (0), Z *or contain* [R, R]. *The only subalgebras of* R *invariant under all special inner automorphisms of* R *are* (0), Z *and* R.

Proof. Let W be a subspace invariant under all special inner automorphisms of R. Given $a \in R$ with $a^2 = 0$ and $w \in W$ then $(1+a)w(1-a) = w + aw - wa - awa \in W$, hence $aw - wa - awa \in W$. Let $\alpha \in F$ with $\alpha^2 \neq \alpha$; since $(\alpha a)^2 = 0$ we get $\alpha aw - w(\alpha a) - (\alpha aw(\alpha a) \in W$, that is, $\alpha(aw - wa) - \alpha^2 awa \in W$. Eliminating with the above yields $(\alpha^2 - \alpha)(aw - wa) \in W$ and since $\alpha^2 - \alpha \neq 0$ is in F we end up with $aw - wa \in W$.

If e is an idempotent in R then $a = xe - exe$ satisfies $a^2 = 0$ for all $x \in R$ so $(xe - exe)w - w(xe - exe) \in W$. Similarly, $(ex - exe)w - w(ex - exe) \in W$. The net result of these is that $(ex - xe)w - w(ex - xe) \in W$, which is to say, $[[E, R], W] \subset W$. Since $[E, R] \supset [R, R]$ by the corollary to Lemma 1.10, we get $[[R, R], W] \subset W$. Hence if R is not 4-dimensional over a field of characteristic 2, by Theorem 1.14 we get $W \subset Z$ or $W \supset [R, R]$. Since W is a subspace over the centroid, $W \subset Z$ implies $W = (0)$ or $W = Z$.

So if R is not 4-dimensional over Z of characteristic 2 we are done. Suppose that R is 4-dimensional over Z; since R has a non-trivial idempotent, R is not a division ring so $R \approx Z_2$ the 2 by 2 matrices over Z, $Z \neq GF(2)$ (and $F = Z$ in this case), and $W \subset Z_2$ is invariant under all inner automorphisms of Z_2. Given $a = \begin{pmatrix} \alpha & \beta \\ \gamma & \delta \end{pmatrix} \in W$ then $e_{12}a + ae_{12} \in W$ and $a + e_{12}a + ae_{12} + e_{12}ae_{12} \in W$ so that $e_{12}ae_{12} = \begin{pmatrix} 0 & \gamma \\ 0 & 0 \end{pmatrix} \in W$. If $\gamma \neq 0$ we get $e_{12} \in W$ from which $e_{21} \in W$ follows by applying the inner automorphism induced by $\begin{pmatrix} 0 & 1 \\ 1 & 0 \end{pmatrix}$, and so $e_{11} + e_{22} = e_{12}e_{21} + e_{21}e_{12} \in W$. But then $W \supset [R, R]$. So if $W \not\supset [R, R]$ we must assume $\begin{pmatrix} \alpha & \beta \\ \gamma & \delta \end{pmatrix} \in W$ forces $\gamma = 0$, and similarly $\beta = 0$. That is, W consists of diagonal matrices. If $\begin{pmatrix} \alpha & 0 \\ 0 & \delta \end{pmatrix} \in W$, with $\alpha \neq \delta$ then

$$\begin{pmatrix} 1 & 1 \\ 0 & 1 \end{pmatrix}\begin{pmatrix} \alpha & 0 \\ 0 & \delta \end{pmatrix}\begin{pmatrix} 1 & 1 \\ 0 & 1 \end{pmatrix} = \begin{pmatrix} \alpha & \delta \\ 0 & \delta \end{pmatrix}\begin{pmatrix} 1 & 1 \\ 0 & 1 \end{pmatrix} = \begin{pmatrix} \alpha & \alpha+\delta \\ 0 & \delta \end{pmatrix}$$

is in W, forcing $\alpha + \delta = 0$, that is $\alpha = \delta$. Thus if $W \not\supset [R, R]$ then $W = Z$.

If W is, in addition, a subring of R being (0), Z or containing $[R, R]$, in light of the corollary to Theorem 1.5, W is (0), Z or R.

This theorem has many applications of a certain form; for instance, let M be the additive subgroup generated by all elements of square 0; then M is invariant under all automorphisms so in the presence of an idempotent $M \supset [R, R]$. Similarly for sums of invertible elements, quasi-regular elements, etc.

If one wanted to study the existence or nature of a simple nil ring -- and in light of Golod's recent results this looks likely -- at least when this simple ring has an infinite centroid one can easily show that the subring generated by the elements of square 0 is a Lie ideal of R, so must be R itself. Thus to investigate local finiteness of R one could content oneself with studying the subring generated by a finite number of elements whose square is 0.

Let's return to Theorem 1.15 for a moment. If R is simple without a non-trivial idempotent the result may be false as we show by an example.

Let F be a field of characteristic 0 and $'$ a non-trivial derivation of F into itself, that is $(\alpha + \beta)' = \alpha' + \beta'$, $(\alpha\beta)' = \alpha'\beta + \alpha\beta'$. Consider all polynomials $F[t]$ in t where we suppose that for $\alpha \in F$, $\alpha t - t\alpha = \alpha'$. It is easy to show that $F[t]$ is a simple, non-commutative integral domain. Now the only invertible elements in $F[t]$ are precisely the non-zero elements of F so that F is invariant under all automorphisms of $F[t]$. Yet $F \not\subset Z$ nor does $F = F[t]$.

One might ask: can we weaken the hypothesis of Theorem 1.15 as follows; let R be a simple ring with zero divisors, then is a subring invariant under all automorphisms in the center or all of the ring? Sasiada, in an unpublished result, has shown that simple rings with zero divisors can have proper subrings, not in the center, invariant under all automorphisms. In fact in his example such a subring can be chosen to consist of all the nilpotent elements in the ring.

Bibliography

1. S.A. Amitsur, "Derivations in simple rings," Proc. London Math. Soc. 7 (1957), pp. 87-112.

2. S.A. Amitsur, "Invariant submodules of simple rings," Proc. Amer. Math. Soc. 7 (1958), pp 987-989.

3. W. Baxter, "Lie simplicity of a special class of associative rings," Proc. Amer. Math. Soc. 7 (1958), pp. 855-863.

4. I.N. Herstein, "On the Lie and Jordan rings of a simple associative ring," Amer. Journal of Math. 77 (1955), pp. 279-285.

5. I.N. Herstein, "The Lie ring of a simple associative ring," Duke Math. Journal 22 (1955), pp. 471-476.

6. I.N. Herstein, "Lie and Jordan structures in simple associative rings," Bull. Amer. Math. Soc. 67 (1961), pp. 517-531.

CHAPTER 2

SIMPLE RINGS WITH INVOLUTION

Let R be a simple ring in which there is defined an involution * satisfying $a^{**} = a$, $(a+b)^* = a^* + b^*$, $(ab)^* = b^*a^*$. Let $S = \{x \in R | x^* = x\}$ and $K = \{x \in R | x^* = -x\}$.

It is immediately apparent that there is a sharp difference between the case in which R is of characteristic 2 and that in which its characteristic is not 2. For the most part we shall content ourselves with a study of the latter case.

Suppose then that, in this chapter, unless otherwise specified R will denote a simple ring with involution of characteristic different from 2. In this case, since 2R is an ideal of R so must be R, for any $x \in R$ $x/2$ makes sense and so, since

$$x = \frac{x + x^*}{2} + \frac{x - x^*}{2}, \quad R = S + K, \quad S \cap K = (0).$$

S is a Jordan subring of R and K a Lie subring of R; we shall determine the nature of S as a Jordan ring and that of K and [K, K] as Lie rings.

In studying rings with involution two cases immediately present themselves; these depend on the nature of the involution on a certain prescribed subset. The definition we are about to give should be made using the centroid rather than the center, however in the material at hand it is the center, even if it is 0, that plays the crucial role.

DEFINITION. The involution is of the <u>first kind</u> if $Z \subset S$; the involution is of the <u>second kind</u> if $Z \not\subset S$.

A study of the Lie and Jordan structure of simple rings with involutions of the second kind is easy and we shall dispose of these soon. Involutions of the first kind offer much deeper difficulties.

Let us recall at this point that in Theorem 1.6 we showed that in a simple ring with involution whose dimension over its center exceeds 4 the subring generated by the symmetric elements is the ring itself.

<u>Notation.</u> If A is a subset of R then $\overline{A}$ will denote the subring of R generated by A.

Let R have an involution of the second kind (we do not restrict the characteristic of R), and let $F = \{a \in Z \mid a^* = a\}$. Then $[Z:F] = 2$ and it is easy to verify that $R \otimes_F Z$ is a direct sum of two simple rings and that the involution on $R \otimes_F Z$ interchanges these simple rings. The Lie and Jordan results in S and K can be obtained from the results developed in Chapter 1 on simple rings without involution. However we choose to follow a different path, working in R itself.

Let R be simple, of arbitrary characteristic having an involution of the second kind. Let $Z_S = Z \cap S$. Then it follows that $[Z:Z_S] = 2$ and $Z = Z_S(\alpha)$. From this it is immediate that $R = S + \alpha S$.

If the characteristic of R is not 2 let U be a Jordan ideal of S; then we claim that $U + ZU$ is a Jordan ideal of R, as can immediately be verified, so that if $U \neq (0)$ then $U + ZU$ is R. But then $Z_S U = S$, on comparing R with $S + ZS$, hence $U \supset S \circ U = (Z_S S) \circ U = S \circ (Z_S U) = S \circ S$, that is, $U \supset S \circ S$. Since

$R = R\circ R$ (R is simple as a Jordan ring) then $S\circ S + \alpha S\circ S = R\circ R = R$, we get $U = S$.

If, in the case of arbitrary characteristic for R, V is a Lie ideal of $[K, K]$ then, as can be verified, since $[R, R] = [K, K] + \alpha[K, K]$, and since $V + ZV$ is a Lie ideal of $[R, R]$, by Theorem 1.14, if R is not 4-dimensional over Z, Z of characteristic 2, then either $V + ZV \subset Z$ or $V + ZV = [R, R]$.

The first possibility implies that $V \subset Z$, the second that $V + \alpha Z_S V = [K, K]$. Since $[[R, R], [R, R]] = [R, R]$ it follows that $[[K, K], [K, K]] = [K, K]$ and so $[V + \alpha Z_S V, V + \alpha Z_S V] = [K, K]$. Thus $[V, V] + [V, \alpha Z_S V] + [\alpha Z_S V, \alpha Z_S V] \supset [K, K]$, and so $V \supset [V, [K, K]] \supset [V, V] + [V, \alpha Z_S V] + [V, \alpha^2 Z_S V] \supset [V, V] + [V, \alpha Z_S V] + [\alpha Z_S V, \alpha Z_S V] \supset [K, K]$.

We have proved

THEOREM 2.1. *Let* R *be a simple ring having an involution of the second kind. Then*

1. *if the characteristic of* R *is not* 2 *then* S *is a simple Jordan ring.*

2. *any Lie ideal of* $[K, K]$ *is either in* Z *or is* $[K, K]$ *except if* R *is of characteristic* 2 *and is* 4-*dimensional over its center*.

3. *any Lie ideal of* K *either is in* Z *or contains* $[K, K]$ *except if* R *is of characteristic* 2 *and is* 4-*dimensional over its center.*

We leave the proof of part 3 to the reader.

With this theorem out of the way we can focus our attention on the case of involutions of the first kind. While many of the results to follow hold irrespective of the nature of the involution we henceforth assume in this chapter that R *is a simple ring of characteristic not 2 having an involution of the first kind.*

We have already seen in Theorem 1.6 that $\overline{S} = R$ for most simple rings. We now wish to establish its companion theorem, namely, that $\overline{K} = R$ in general. To do so we first show another construction, in most rings with involution, of a Lie ideal of the ring.

DEFINITION. K^2 is the additive group generated by all k_1k_2, with $k_1, k_2 \in K$.

LEMMA 2.1. _Let_ R _be any ring with involution such that_ $R = S + K$. _Then_ K^2 _is a Lie ideal of_ R.

Proof. Let $k_1, k_2 \in K$ and $k \in K$. Then $(k_1k_2)k - k(k_1k_2) = (k_1k - kk_1)k_2 + k_1(k_2k - kk_2) \in K^2$, so $[K^2, K] \subset K^2$. On the other hand, if $s \in S$ then $(k_1k_2)s - s(k_1k_2) = k_1(k_2s + sk_2) - (k_1s + sk_1)k_2$ which is in K^2 since $K \circ S \subset K$. Thus $[K^2, S] \subset K^2$. Since $R = S + K$ we have $[K^2, R] \subset K^2$ whence K^2 is a Lie ideal of R.

The lemma enables us to prove

THEOREM 2.2. _If_ R _is a simple ring with involution, of characteristic not_ 2, _then_ $\overline{K} = R$ _provided_ $\dim R/Z > 4$.

Proof. The conditions of Lemma 2.1 hold in R hence K^2, as a Lie ideal of R, by Theorem 1.3 must either contain $[R, R]$ or $K^2 \subset Z$.

Now if $K^2 \supset [R, R]$ then $\overline{K} \supset \overline{K^2} \supset \overline{[R, R]} = R$ by the corollary to Theorem 1.3. Suppose then that $K^2 \subset Z$.

If $a \in K$ is not invertible then since $aK \subset Z$ and all the non-zero elements of Z are invertible we must have $aK = (0)$. In particular, $a^2 = 0$. If $s \in S$ then $as + sa \in K$ hence $0 = a(as + sa) = a^2s + asa = asa$; therefore $aSa = (0)$. Together with $aK = (0)$ we get $aRa = (0)$; consequently Ra is a nilpotent left ideal and so $a = 0$. Thus $a \neq 0$ in K forces a to be invertible; if $b \in K$ then since $ab \in Z$ we get $b \in Za^{-1} = Za$ (since $a^2 = \sigma \in Z$) which is to say, $K \subset Za$.

If $s \in S$ commutes with a then $sa \in K \subset Za$ forcing $s \in Z$. Now if $s \in S$ then $sa + as = \lambda a$, $\lambda \in Z$, in fact $\lambda \in Z \cap S$. Thus $(s^2 - \lambda s)a = a(s^2 - \lambda s)$; but since $s^2 - \lambda s \in S$ and commutes with a, by the above remark, $s^2 - \lambda s \in Z$.

Given $x \in R$, $x = s + \alpha a$, $s \in S$, $\alpha \in Z$. Hence $x^2 = s^2 + \alpha(sa + as) + \alpha^2 a^2 = s^2 + \lambda\alpha a + \alpha^2\sigma$ where $as + sa = \lambda a$. Thus $x^2 - \lambda x = s^2 - \lambda s + \alpha^2\sigma$ and since, as we have seen, $s^2 - \lambda s \in Z$ we must have $x^2 - \lambda x \in Z$. In this way R has been shown to be quadratic over Z. By Theorem 2.38 of (N), R must be at most 4-dimensional over Z. This proves the theorem.

We sharpen this result to get an extension of a result of Baxter.

THEOREM 2.3. Let R be a ring with involution such that $2R = R$ in which $\overline{K} = R$. Then $S = K \circ K$; that is, given $s \in S$ then $s = \sum a_i^2 - \sum b_i^2$ where $a_i, b_i \in K$.

Proof. The hypothesis implies that $2S = S$, $2K = K$ and $R = S \oplus K$.

Let $k_1, k_2 \in K$; then $k_1k_2 - k_2k_1 \in K$ and $k_1k_2 + k_2k_1 \in K \circ K$ so $2k_1k_2 = (k_1k_2 - k_2k_1) + (k_1k_2 + k_2k_1) \in K + K \circ K$. This gives $K^2 \subset K + K \circ K$. We want to show that $K^n \subset K + K \circ K$ for any positive integer n, so we go by induction on n assuming that $K^{n-1} \subset K + K \circ K$ for $n \geq 3$.

Given $k_1, \ldots, k_n \in K$ then $(k_1k_2 - k_2k_1)k_3 \ldots k_n \in K^{n-1} \subset K + K \circ K$ as is $k_2(k_1k_3 - k_3k_1)k_4 \ldots k_n \in K + K \circ K$. Adding we have $(k_1k_2k_3 - k_2k_3k_1)k_4 \ldots k_n \in K + K \circ K$. However, since $k_1k_3k_2 + k_2k_3k_1 \in K$, $(k_1k_3k_2 + k_2k_3k_1)k_4 \ldots k_n \in K^{n-2} \subset K + K \circ K$; adding this to $(k_1k_2k_3 - k_2k_3k_1)k_4 \ldots k_n$ which we have seen is in

$K + K\circ K$ we obtain $k_1(k_2k_3 + k_3k_2)k_4 \ldots k_n \in K + K\circ K$. Since $k_1(k_2k_3 - k_3k_2)k_4 \ldots k_n \in K^{n-1} \subset K + K\circ K$ we get $2k_1k_2 \ldots k_n \in K + K\circ K$ so that $K^n \subset K + K\circ K$, completing the induction. Since $\sum K^n = \overline{K} = R$ we get $R = K + K\circ K$; clearly $K\circ K \subset S$. Since $R = K + S$, $K \cap S = (0)$ we conclude that $S = K\circ K$.

Given $s \in S$, $s = \sum (a_ib_i + b_ia_i)$ with $a_i, b_i \in K$ hence $s = \sum (a_i + b_i)^2 - \sum a_i^2 - \sum b_i^2$. This proves the theorem completely.

We return to simple rings; for these this theorem has a very interesting consequence, namely

THEOREM 2.4. *If* K *has only a finite number of elements linearly independent over* Z *then* R *is finite-dimensional over* Z; *in fact* $\dim R/Z \leq n^2 + n$, *where* n *is the maximal number of independent elements in* K.

Proof. If $k_1, \ldots, k_n$ is a maximally independent set of elements in K over Z then since $S = K\circ K$ we have that $k_ik_j + k_jk_i$ span S over Z. Thus $\dim R/Z \leq n^2 + n$.

We now prepare to study the Jordan structure of S. We begin with

THEOREM 2.5. *If* $U \neq (0)$ *is a Jordan ideal of* S *then for* $u \in U$, $r, s \in R$, $ru^4s + s^*u^4r^* \in U$.

Proof. The proof will consist of breaking r and s into their symmetric and skew parts and verifying that in these special instances the theorem holds. We do this in the sequence of three lemmas.

LEMMA 2.2. *If* $x, y \in S$ *and* $u \in U$ *then* $xuy + yux \in U$.

Proof. $2xux = \{x(xu + ux) + (xu + ux)x\} - \{x^2u + ux^2\}$. Since $x^2 \in S$, $x^2u + ux^2 \in U$; since $xu + ux \in U$ so is $x(xu + ux) + (xu + ux)x$. Thus $2xux \in U$; since $2S = S$ we get $xux \in U$. Linearizing on x we obtain the lemma.

LEMMA 2.3. *If* $s \in S$ *and* $k \in K$ *and* $u \in U$ *then* $su^2k - ku^2s \in U$.

Proof. We notice, first of all, that since $uk - ku$ is in S, $u^2k - ku^2 = u(uk - ku) + (uk - ku)u$ is in U. Being a Jordan ideal of S, $s(u^2k - ku^2) + (u^2k - ku^2)s \in U$. That is

(1) $\quad s(u^2k - ku^2) + (u^2k - ku^2)s = su^2k - ku^2s - sku^2 + u^2ks$

is in U.

Consider

(2) $\quad k(u^2s - su^2) + (u^2s - su^2)k = ku^2s - su^2k - ksu^2 + u^2sk$.

Adding (1) and (2) the right sides add up to $u^2(ks + sk) - (ks + sk)u^2$ which, since $ks + sk \in K$, we have seen must be in U. Therefore the sum of the left sides must be in U; since the left side of (1) is already in U we get that of (2) must also be. Now subtract (1) from (2); doing so we stay in U. The result on the right is $2(su^2k - ku^2s) \in U$. Since $2S = S$ this gives $su^2k - ku^2s \in U$ for all $s \in S$, $k \in K$, $u \in U$, which is the desired result.

LEMMA 2.4. *If* $a, b \in K$ *and* $u \in U$ *then* $au^4b + bu^4a \in U$.

Proof. Since $b \in K$ we have already seen that $bu^2 - u^2b \in U$ so that $2(bu^2 - u^2b)^2 \in U$. Since $2K = K$ (and so $4K = K$) this gives us $(bu^2 - u^2b)^2 \in U$. But expanding we have $(bu^2 - u^2b)^2 = \{(bu^2b)u^2 + u^2(bu^2b)\} - u^2b^2u^2 - bu^4b$. Now the $\{\quad\}$ is in U since $bu^2b \in S$ and $2u^2 \in U$. By Lemma 2.2, since $2u^2 \in U$, $4u^2b^2u^2 \in U$, whence $u^2b^2u^2 \in U$. The upshot of all this is that $bu^4b \in U$. Linearizing on b we get the lemma.

Proof of Theorem 2.5. Given $u \in U$, $r, s \in R$ then $r = r_0 + r_1$, $s = s_0 + s_1$ with $r_0, s_0 \in S$, $r_1, s_1 \in K$. Thus

$$ru^4s + s^*u^4r^* = (r_0 + r_1)u^4(s_0 + s_1) + (s_0 - s_1)u^4(r_0 - r_1)$$
$$= (r_0u^4s_0 + s_0u^4r_0) + (r_0u^4s_1 - s_1u^4r_0)$$
$$+ (r_1u^4s_0 - s_0u^4r_1) + (r_1u^4s_1 + s_1u^4r_1).$$

Since $4u^4 \in U$ and since we have seen the factor 4 can be eliminated we obtain the desired theorem as a combination of the previous three lemmas.

We are in a position to prove the basic

THEOREM 2.6. The only Jordan ideals of S are (0) and S; that is, S is a simple Jordan ring.

Proof. Let $U \neq (0)$ be a Jordan ideal of S. If $u \in U$ then we have seen that $ru^4t + t^*u^4r^* \in U$ for all $r, t \in R$. If $u^4 \neq 0$ then $Ru^4R = R$ and so, given $x \in R$ then $x = \Sigma r_iu^4t_i$. But then $x^* = \Sigma t_i^*u^4r_i^*$ leading to $x + x^* = \Sigma (r_iu^4t_i + t_i^*u^4r_i^*)$ so is in U. Since $x + x^*$ covers S as x runs over R we get that $U = S$.

Thus if $U \neq S$ we must assume that $u^4 = 0$ for all $u \in U$. Given $u \in U$, $r = r_0 + r_1$, $r_0 \in S$, $r_1 \in K$ we have $2(u^2r + r^*u^2) = 2(u^2r_0 + r_0u^2) + 2(u^2r_1 - r_1u^2)$ is in U; but since $(u^2)^2 = 0$, $4(u^2r + r^*u^2)^4 = 0$ we get $(u^2r)^5 = 0$ for all $u \in U$ and $r \in R$. By Lemma 1.1 we conclude that $u^2 = 0$ for all $u \in U$. Linearizing we get that $uv + vu = 0$ for all $u, v \in U$. Given $s \in S$, $v = us + su \in U$ hence $0 = u(us + su) + (us + su)u = 2usu$ since $u^2 = 0$. Thus $usu = 0$ for all $u \in U$, $s \in S$. Given $k \in K$ then $kuk \in S$ hence $ukuku = 0$. For any $r \in R$, $r = r_0 + r_1$ with $r_0 \in S$, $r_1 \in K$, whence $ururu = u(r_0 + r_1)u(r_0 + r_1)u = ur_1ur_1u = 0$. That is $(ur)^3 = 0$ for all $r \in R$. By Lemma 1.1 we conclude that $u = 0$. We have proved that $U = (0)$ or $U = S$; this establishes the theorem.

Having determined the Jordan structure of S we now want to determine the Lie structure of K and of [K, K]. We begin with the very easy

LEMMA 2.5. _If_ U _is a Lie ideal of_ K _and if_ $u \in U$, $s \in S$ _then_ $u^2 s - su^2 \in U$.

Proof. To see the result merely note that $us + su \in K$ and $u^2 s - su^2 = u(us + su) - (us + su)u$.

Motivated by what we did in Chapter 1 we make the

DEFINITION. If U is a Lie ideal of K then $T(U) = \{x \in K | [x, K] \subset U\}$.

Clearly T(U) is a Lie ideal of K and contains U. We want a closer tie-in between U and T(U). This is given us by

LEMMA 2.6. _If_ U _is a Lie ideal of_ K _then_ $u, v, w \in U$ _implies_ $uvu \in T(U)$ _and_ $uvw + wvu \in T(U)$.

Proof. Consider [uvu, K]; for $k \in K$

$$uvuk - kuvu = u(vuk + kuv) - (vuk + kuv)u + vuku - ukuv.$$

Since $vuk + kuv = vuk - (vuk)^*$, it is in K so its commutator with u is in U. Since $uku \in K$, $v(uku) - (uku)v \in U$. In all we have shown that $[uvu, K] \subset U$ and so $uvu \in T(U)$. Linearizing on U we obtain $uvw + wvu \in T(U)$.

We proceed to prove

THEOREM 2.7. _If_ U _is a Lie ideal of_ K _then for all_ $u, v \in U$, $(u^2 v - vu^2)r - r^*(u^2 v - vu^2) \in T(U)$ _for all_ $r \in R$.

Proof. We write $r = s + k$ with $s \in S$, $k \in K$, and verify the theorem for s and k.

1. $(u^2v - vu^2)s - s(u^2v - vu^2) = \{(u^2s - su^2)v - v(u^2s - su^2)\}$
$$+ \{(u^2(vs - sv) - (vs - sv)u^2\}.$$

However, by Lemma 2.5, $u^2s - su^2 \in U$, so the first $\{\ \}$ is in U. Also, since $vs - sv \in S$, by Lemma 2.5 again the second $\{\ \}$ is in U. Being in U these are certainly in $T(U)$.

2. $(u^2v - vu^2)k + k(u^2v - vu^2)$

$$= \{u^2(vk + kv) - (vk + kv)u^2\} + v(ku^2 - u^2k) + (ku^2 - u^2k)v$$

$$\equiv v(ku^2 - u^2k) + (ku^2 - u^2k)v \mod U \text{ (by Lemma 2.5 since } vk + kv \in S)$$

$$\equiv v((ku - uk)u + u(ku - uk)) + ((ku - uk)u + u(ku - uk))v \mod U$$

$$\equiv v\{(ku - uk)u - u(ku - uk)\} - \{(ku - uk)u - u(ku - uk)\}v + 2(vu(ku - uk) + (ku - uk)uv) \mod U$$

$$\equiv 2(vu(ku - uk) + (ku - uk)uv) \mod U.$$

But by Lemma 2.5, since $ku - uk \in U$, $vu(uk - ku) + (uk - ku)uv$ is in $T(U)$. The upshot of all this is that $(u^2v - vu^2)k + k(u^2v - vu^2) \in T(U)$. Thus we indeed have proved the assertion of the theorem.

This theorem enables us to establish the key fact contained in

THEOREM 2.8. <u>If</u> R <u>is simple and</u> $\dim R/Z > 4$ <u>and if</u> U <u>is a Lie ideal of</u> K <u>then either</u> $U \supset [K, K]$ <u>or</u> $u^2v = vu^2$ <u>for all</u> $u, v \in U$.

<u>Proof</u>. Let $a = u^2v - vu^2$ where $u, v \in U$. By the previous theorem $ar - r^*a \in T(U)$ for all $r \in R$.

If $k_1 \in K$ then $b = (ar - r^*a)k_1 - k_1(ar - r^*a) \in U \subset T(U)$. Since $b = a(rk_1) - (rk_1)^*a - r^*ak_1 - k_1ar$ we end up with $k_1ar + r^*ak_1 \in T(U)$ for all $r \in R$, $k_1 \in K$.

We continue in this vein; let $k_2 \in K$. Then

$$(r^*ak_1 + k_1ar)k_2 - k_2(r^*ak_1 + k_1ar) \in T \subset T(U),$$

hence

$$r^*ak_1k_2 - (r^*ak_1k_2)^* + k_1a(rk_2) + (rk_2)^*ak_1 \in T(U) ,$$

and since $k_1a(rk_2) + (rk_2)^*ak_1 \in T(U)$, we obtain $r^*ak_1k_2 - (r^*ak_1k_2)^* \in T(U)$. Continuing we get, by induction that for all $\overline{k} \in \overline{K}$, $r^*a\overline{k} - (r^*a\overline{k})^* \in T(U)$.

Since $\dim R/Z > 4$, by Theorem 2.2, $\overline{K} = R$; the above then translates to: for all $r, t \in R$, $rat - (rat)^* \in T(U)$. Now RaR is an ideal of R, hence, if $a \neq 0$, $RaR = R$. But then for any $x \in R$, $x = \Sigma\, r_iat_i$, whence $x^* = \Sigma\, t_i^*ar_i^*$ and so $x - x^* = \Sigma\,(r_iat_i - (r_iat_i)^*) \in T(U)$. Since $x - x^*$ sweeps out K we have that if $a \neq 0$ then $T(U) \supset K$. From the definition of $T(U)$ this says that $U \supset [K, K]$. The theorem has now been proved.

THEOREM 2.9. *If* R *is simple,* $\dim R/Z > 4$ *and if* U *is a Lie ideal of* K *such that* $U \not\supset [K, K]$ *then given* $u \in U$, $u^2 \in Z$.

Proof. Since $u^2v = vu^2$ by the previous theorem, u^2 is in the center of $\overline{U}$, the subring generated by U. However, $u^2s - su^2 \in U$ for $s \in S$ and $u^2k - ku^2 = u(uk - ku) + (uk - ku)u \in \overline{U}$ for $k \in S$, thus u^2 commutes with all $u^2(s+k) - (s+k)u^2$, $s \in S$, $k \in K$, that is, with all $u^2r - ru^2$, $r \in R$. Since the characteristic is not 2 we have seen in the sublemma in Chapter 1 that this forces u^2 to be in Z.

COROLLARY. *If* $U \not\supset [K, K]$ *then* $uv + vu \in Z$ *for all* $u, v \in U$.

One could ask: we know $u^2 \in Z$ for all $u \in U$; is it possible that $u^2 = 0$ for all these u's? This is answered in the

THEOREM 2.10. *If* R *is simple*, dim $R/Z > 4$ *and* U *is a Lie ideal of* K *such that* $u^2 \in U$ *implies* $u^2 = 0$ *then* $U = (0)$.

Proof. On linearizing $u^2 = 0$ we get $uv + vu = 0$ for all $u, v \in U$. Thus $uvu = -vu^2 = 0$ for any $u, v \in U$. Given $u \in U$, $k \in K$ then $2uku = (uk - ku)u - u(uk - ku)$ (since $u^2 = 0$), hence $uku \in U$; but then $vukuv = 0$. Since $uv = -vu$, we arrive at $uvKuv = (0)$. Now $(uv)^* = v^*u^* = vu = -uv$, that is, $uv \in K$, thus for $s \in S$, $suvs \in K$ and so $uvsuvsuv \in uvKuv = (0)$. Given $r \in S$, $r = s + k$, $s \in S$, $k \in K$, whence $uvruvruv = 0$. The right ideal uvR is such that every element in it has cube 0. By Lemma 1.1 this forces $uv = 0$ for all $u, v \in U$. But then for $k \in K$, $u(uk - ku) = 0$, leaving us with $uKu = (0)$. As above we then get $ususu = 0$ for $s \in S$ and so uR is a nil right ideal where every element has cube 0. The outcome of this is that $u = 0$, that is, $U = (0)$.

Combining Theorems 2.9 and 2.10 we have

THEOREM 2.11. *If* R *is simple and* $Z = (0)$ *then any non-zero Lie ideal* U *of* K *must contain* $[K, K]$.

We henceforth may assume that $Z \neq (0)$. We recall the precise situation we are studying. R is simple, of characteristic $\neq 2$, $Z \neq (0)$, dim $R/Z > 4$ and R has an involution of the first kind.

Let U be a Lie ideal of K such that $U \not\supset [K, K]$. According to Theorems 2.9 and 2.10 every element $w \in U$ satisfies $w^2 \in Z$ and, furthermore, there is an element $u \in U$ such that $u^2 = \lambda_1 \neq 0 \in Z$. Since K generates R and $K \cap Z = (0)$, $u \notin Z$ so u can not commute with every element $k \in K$. We claim something more, namely that there is a $k \in K$ such that $(uk - ku)^2 \neq 0$.

Suppose, then, that $(uk - ku)^2 = 0$ for all $k \in K$; linearizing we obtain

(1) $$(uk_1 - k_1u)(uk_2 - k_2u) + (uk_2 - k_2u)(uk_1 - k_1u) = 0$$

for all $k_1, k_2 \in K$.

Since $u^2 \in Z$ each $uk_i - k_iu$ anti-commutes with u hence $(uk_1 - k_1u)(uk_2 - k_2u) - (uk_2 - k_2u)(uk_1 - k_1u)$ is both in U and commutes with u; by (1) this element equals $w = 2(uk_1 - k_1u)(uk_2 - k_2u)$. Since it commutes with u and $uw + wu \in Z$ we get $2uw \in Z$; but w is not invertible since $uk_2 - k_2u$ is not, so $2uw$ is not, hence $2uw = 0$. Because $2u$ is invertible we end up with $w = 0$.

Let $v = uk_1 - k_1u \neq 0$; then, by the above, $v(uk - ku) = 0$ for all $k \in K$, that is, $vuk = vku$. Now $kuk \in K$ hence $vukuk = vkuku = vuk^2u$ from the above. Also $vukuk = vku^2k = vk^2u^2$; comparing we have $vk^2u^2 = vuk^2u$. But u is invertible hence $v(k^2u - uk^2) = 0$. By Theorem 2.3 the k^2 additively span S so $v(us - su) = 0$ for all $s \in S$. But then $v(ur - ru) = 0$ for all $r \in R$. Given $r, a \in R$, $0 = v(ura - rau) = v(ur - ru)a + vr(ua - au) = vr(ua - au)$, which is to say, $vR(ua - au) = (0)$. Since R is prime and $v \neq 0$ we get $ua = au$ for all $a \in R$, therefore $u \in Z$ contrary to $u \notin Z$. We have substantiated our claim that there is a $v = uk - ku$ so $v^2 = \lambda_2 \neq 0 \in Z$. Notice that $uv + vu = 0$ and $v \in U$. A calculation reveals that $(uv - vu)^2 = -4\lambda_1\lambda_2 \neq 0$.

We claim that $u, v, uv - vu$ are linearly independent over Z, for if $w = \alpha_o u + \alpha_1 v + \alpha_2(uv - vu) = 0$, $\alpha_i \in Z$ then $0 = uw + wu = 2\alpha_o\lambda_1$, $0 = vw + wv = 2\alpha_1\lambda_2$ so that $\alpha_o = \alpha_1 = 0$, forcing $\alpha_2 = 0$ in addition.

We further claim that $u, v, uv - vu$ form a basis of U over Z. For if there is an $x \in U$ linearly independent of these, since $ux + xu = \alpha$, $vx + xv = \beta$, $\alpha, \beta \in Z$, letting $x' = x + \sigma_1 u + \sigma_2 v$,

$\sigma_1, \sigma_2 \in Z$ a computation shows that $x'u + ux' = \alpha + 2\sigma_1\lambda_1$, $x'v + vx' = \beta + 2\sigma_2\lambda_2$. Since $\lambda_1 \neq 0$, $\lambda_2 \neq 0$ we can find σ_1, σ_2 such that $x'u + ux' = x'v + vx' = 0$; moreover x' is linearly independent of $u, v, uv - vu$. We showed that without loss of generality there is an $x \in U$ linearly independent of $u, v, uv - vu$ which satisfies $xu + ux = xv + vx = 0$. Now $x(uv - vu) = (uv - vu)x$ which together with $x(uv - vu) + (uv - vu)x \in Z$ forces $2x(uv - vu) = \gamma \in Z$. Since $(uv - vu)^2 = -4\lambda_1\lambda_2 \in Z$ we get $-8\lambda_1\lambda_2 x = \gamma(uv - vu)$ forcing $x = \delta(uv - vu)$ with $\delta \in Z$, a contradiction.

We have shown that any Lie ideal $U \neq (0)$ of K which does not contain $[K, K]$ must be 3-dimensional over Z; in fact U must contain a basis of the form $u, v, uv - vu$ where $u^2 = \lambda_1 \neq 0$, $v^2 = \lambda_2 \neq 0$, $uv + vu = 0$ and where $\lambda_1, \lambda_2 \in Z$.

Let $W = \{x \in K \mid xu = ux, xv = vx\}$. W clearly is $\{x \in K \mid [x, U] = (0)\}$. The Jacobi identity reveals that W is a Lie ideal of K. Moreover, since $uv - vu \notin W$ and is in $[K, K]$ we must have $W \not\supset [K, K]$. From the above discussion either $W = (0)$ or W itself is 3-dimensional over Z. Knowing the form of U we readily check that $U \cap W = (0)$.

Let $s \in S$; we claim that $a = (us - su)v + v(us - su)$ is in W. Clearly, at any rate, it is in K; moreover, since $v^2 \in Z$ a commutes with v. But $(us - su)v + v(us - su) = u(sv - vs) + (sv - vs)u$ since $uv = -vu$, hence a also commutes with u. Consequently $a \in W$.

If $W = (0)$ we would have that $(us - su)v + v(us - su) = 0$ for all $s \in S$. Now $t = vs - sv \in S$ therefore $(ut - tu)v + v(ut - tu) = 0$. Since t and u anti-commute with v, $ut - tu$ commutes with v, thus we get $2(ut - tu)v = 0$. But v is invertible, reducing this to $ut - tu = 0$. However, $ut + tu = u(vs - sv) + (vs - sv)u = 0$, thus

$2ut = 0$ and so since u is invertible $t = 0$. That is, $v \in U$ centralizes all of S. But $\overline{S} = R$ (Theorem 1.6) hence $v \in Z$, a contradiction since $uv - vu \neq 0$. Thus $W \neq (0)$. By the argument already given W must be 3-dimensional over Z.

We claim that $W \subset [K, K]$. Clearly, $[W, K] \subset W$ is contained properly in $[K, K]$ and is a Lie ideal, so is 3-dimensional over Z or is (0). If (0), then W centralizes $\overline{K} = R$ so is in Z, contrary to $W \cap K = (0)$. Thus $[W, K]$ is 3-dimensional, so must be all of W.

Now $W \cap U = (0)$ and $W + U \subset [K, K]$ is a 6-dimensional Lie ideal of K. By the above, not being 3-dimensional, $W + U = [K, K]$.

We claim that $[K, K] = K$. If not, there is an $s \in S$ such that $us + su \notin [K, K]$. For if $us + su \in [K, K]$ for all $s \in S$, since $[u, K] \subset [K, K]$ we would have $ur - (ru)^* = ur + r^*u \in [K, K]$ for all $r \in R$. However, we say earlier that this forces $x - x^* \in [K,K]$ for all $x \in R$, that is, $K = [K, K]$. Hence there is an $s_o \in S$ such that $x_1 = us_o + s_ou \notin [K, K]$. Clearly, since $u^2 \in Z$, x_1 commutes with u. Thus $vx_1 - x_1v$ anti-commutes with u. Since $v^2 \in Z$, $vx_1 - x_1v$ anti-commutes with v. But $vx_1 - x_1v \in U$ so $vx_1 - x_1v = \alpha u + \beta v + \gamma(uv - vu)$ with $\alpha, \beta, \gamma \in Z$. Since it anti-commutes with both u and v we readily see that $\alpha = \beta = 0$. Thus $vx_1 - x_1v = \gamma(uv - vu)$ and so $x_1 + \gamma u$ commutes with v; it commutes with u since x_1 does, hence $x_1 + \gamma u \in W \subset [K, K]$. Therefore, since $u \in [K, K]$, $x_1 \in [K, K]$ contrary to assumption. We are thereby forced to conclude that $[K, K] = K$.

Since $K = [K, K] = U + W$ we have that K is 6-dimensional over Z. We know that if v_1, v_2, v_3 is a basis of U and w_1, w_2, w_3 one of W then $v_iv_j + v_jv_i$ and $w_iw_j + w_jw_i$ are in Z for any i,j. Thus $K \circ K$ is spanned by the 10 elements $1, v_iw_j + w_jv_i$, $i, j = 1,2,3$, over Z. By Theorem 2.4, $K \circ K = S$ hence S is at most 10-dimensional over Z. Since K is 6-dimensional over Z and

$R = S + K$ we get $\dim R/Z \leq 16$. In other words, if $\dim R/Z > 16$ and R has an involution of the first kind, then we have proved that any Lie ideal of K must contain $[K, K]$. Knowing the result for involutions of the second kind by Theorem 2.1 we have proved the fundamental

THEOREM 2.12. _If_ R _is a simple ring with involution of characteristic not_ 2 _and if_ U _is a Lie ideal of_ K _then either_ $U \subset Z$ _or_ $U \supset [K, K]$ _provided_ $\dim R/Z > 16$.

We show that the result is false in F_4, F a field, where $*$ is the transpose in F_4. In fact the set

$$U = \left\{ \begin{pmatrix} 0 & \alpha & \beta & \gamma \\ -\alpha & 0 & -\gamma & -\beta \\ -\beta & \gamma & 0 & \alpha \\ -\gamma & \beta & -\alpha & 0 \end{pmatrix} \right\}$$

is a 3-dimensional Lie ideal of K.

We conclude our study of the Lie structure of simple rings by considering the nature of $[K, K]$ as a Lie ring in a simple ring with involution.

In the course of proving Theorem 2.12 we actually proved:

1. If U is a Lie ideal of K such that $u \in U$ implies $u^2 \in Z$ then $U = (0)$ or is 3-dimensional over Z with a basis $u, v, uv - vu$ where $u^2 = \lambda_1 \neq 0$, $v^2 = \lambda_2 \neq 0$, $\lambda_1, \lambda_2 \in Z$ and $uv + vu = 0$.
2. For U as in (1), $W = \{x \in K \mid xa = ax \text{ all } a \in U\}$ is not (0). Since $[W, K] \neq (0)$ (otherwise W would centralize K, hence $\overline{K} = R$ forcing $W \subset Z \cap K = (0)$) there is an element in $[K, K]$ that centralizes U.

From these considerations we prove

THEOREM 2.13. _If_ $\dim R/Z > 4$ _then_ $\overline{[K, K]} = R$.

Proof. Let $u \in [K,K]$; then $u^2s - su^2 \in [K,K]$ and $u^2k - ku^2 \in \overline{[K,K]}$ for $s \in S$, $k \in K$. Thus $u^2r - ru^2 \in \overline{[K,K]}$ for all $r \in R$. But then for $k \in [K,K]$, $\overline{[K,K]} \supset u^2kr - kru^2 = (u^2k - ku^2)r + k(u^2r - ru^2)$. Since $k(u^2r - ru^2) \in \overline{[K,K]}$ we get that $(u^2k - ku^2)R \subset \overline{[K,K]}$ for $u, k \in [K,K]$. Therefore $[K,(u^2k - ku^2)R] \subset [K,\overline{[K,K]}] \subset \overline{[K,K]}$ leaving us with $K(u^2k - ku^2)R \subset \overline{[K,K]}$. Continuing we get $K^n(u^2k - ku^2)R \subset \overline{[K,K]}$; since $\overline{K} = R$ (Theorem 2.2) and $\overline{K} = \Sigma K^n$ we obtain $R(u^2k - ku^2)R \subset \overline{[K,K]}$. If $\overline{[K,K]} \neq R$ we conclude that $u^2k = ku^2$ for $k \in [K,K]$ so u^2 is in the center of $\overline{[K,K]}$. Since $u^2r - ru^2 \in \overline{[K,K]}$, u^2 must commute with all $u^2r - ru^2$, so by the sublemma in Chapter 1, u^2 must be in Z. By the discussion we gave prior to this theorem $[K,K]$ must be 3-dimensional over Z and contains a non-zero element which centralizes it. Since we also have a specific basis $u, v, uv - vu$ of $[K,K]$ over Z where $u^2 = \lambda_1 \neq 0 \in Z$, $v^2 = \lambda_2 \neq 0 \in Z$ and $uv + vu = 0$ we immediately check out that there is no non-zero element in $[K,K]$ which centralizes it. Thus we must conclude that $\overline{[K,K]} = R$.

We assume henceforth that R is more than 16-dimensional over Z.

LEMMA 2.7. *If U is a non-zero Lie ideal of $[K,K]$ then $xU = (0)$ implies $x = 0$.*

Proof. Let $xU = (0)$; for $u \neq 0$ in U and $k \in [K,K]$ we have $0 = x(uk - ku) = -xku$, hence $x[K,K]u = (0)$. Since $x[K,K]U = (0)$, repeating the above argument yields $x[K,K]^2U = (0), \ldots, x[K,K]^nU = (0)$. Since $\overline{[K,K]} = R$ we get $xRU = (0)$. Since $U \neq (0)$ and R is simple this forces $x = 0$.

LEMMA 2.8. *If $U \neq [K,K]$ is a Lie ideal of $[K,K]$ for which $u^2 \in Z$ for all $u \in U$ then $U = (0)$.*

Proof. Suppose that $U \neq (0)$; we claim that there is a $u \in U$ with $u^2 \neq 0$. If not, $uv + vu = 0$ for all $u, v \in U$. Since $uv - vu$ is an element of U we get $2uv \in U$. If $w \in U$ then $2(uvw + wuv) = 0$, hence $uvw = -wuv$. But u, v anti-commute with w therefore uv commutes with w; thus $uvw = 0$. Since $uvU = (0)$ by Lemma 2.7, $uv = 0$. Since $uV = (0)$ we get $u = 0$ contradicting $U \neq (0)$. Therefore there is a $u \in U$ with $u^2 \neq 0$.

Now, for any $v, w \in U$ $vw + wv \in Z$. Also, for $k \in K$, $U \supset v(vk - kv) - (vk - kv)v$; since $v^2 \in Z$, $vk - kv$ anti-commutes with v and so $2v(vk - kv) \in U$. Putting $k = vb - bv$, $b \in K$, we get $v^2(vb - bv) \in U$. Since for some $u \in U$, $u^2 \neq 0 \in Z$ then $u^2K = K$ and since $u^2(ub - bu) \in U$ we see that $ub - bu \in U$ for all $b \in K$. That is, for any $u \in U$ with $u^2 \neq 0$ then $[u, K] \subset U$. If $u^2 = 0$ pick $v \in U$ so that $v^2 \neq 0$. Now not both $(u + v)^2 = 0$ and $(u - v)^2 = 0$ for the two together would imply that $v^2 = 0$; so one of these, say $u+v$ satisfies $(u+v)^2 \neq 0$. But then $[u+v, K] \subset U$; since $v^2 \neq 0$, $[v, K] \subset U$ resulting in $[u, K] \subset U$. All in all we have shown that $[U,K] \subset U$, hence U is a Lie ideal of K. By Theorem 2.12, $U = (0)$.

LEMMA 2.9. <u>If</u> U <u>is a proper Lie ideal of</u> $[K, K]$ <u>then</u> $[U, U] = (0)$ <u>or</u> $\overline{U} = R$.

Proof. Let $a \in [U, U]$; it is immediate that $[a, K] \subset U$. Thus $[a^2, K] = a[a, K] + [a, K]a \subset \overline{U}$. Also $[a^2, S] \subset U$; thus $[a^2, r] \subset \overline{U}$ for all $r \in R$. If $v \in U$ then $a^2vr - vra^2 \in \overline{U}$; but $a^2vr - vra^2 = (a^2v - va^2)r + v(a^2r - ra^2)$. From $v(a^2r - ra^2) \in \overline{U}$ we deduce that $(a^2v - va^2)R \subset \overline{U}$. As before, since $\overline{[K, K]} = R$ we get $R(a^2v - va^2)R \subset \overline{U}$, so, if $\overline{U} \neq R$ then $a^2v = va^2$ for all $v \in U$ from which we see that a^2 centralizes $\overline{U}$. However, $a^2r - ra^2 \in \overline{U}$ for all $r \in R$ so commutes with a^2. From this $a^2 \in Z$ for all $a \in [U, U]$. By Lemma 2.8, $[U, U] = (0)$.

We dispose of the second possibility of Lemma 2.9 by proving

THEOREM 2.14. If U is a proper Lie ideal of $[K, K]$ then $[[U, U],[U, U]] = (0)$.

Proof. We first make the following remark: if $\overline{B} = R$ then $[B, R] = [R, R]$. We prove by induction that $[B^n, R] \subset [B, R]$. If $b_1, \ldots, b_n \in B$, $r \in R$ then

$$b_1 b_2 \ldots b_n r - r b_1 \ldots b_n = b_1(b_2 \ldots b_n r) - (b_2 \ldots b_n r) b_1$$
$$+ b_2 \ldots b_n (r b_1) - (r b_1) b_2 \ldots b_n \in [B, R] + [B^{n-1}, R]$$

is contained in $[B, R]$. Since $R = \Sigma B^n$, $[B, R] = [R, R]$.

Since $\overline{K} = R$, $[K, R] = [R, R]$; from $R = K + S$ this gives us $[K, K] + [K, S] = [K, K] + [S, S] + [K, S]$ so that $[S, S] \subset [K, K]$, and $[R, R] = [K, K] + [K, S]$. Thus if $[[U, U],[U, U]] \neq (0)$ by Lemma 2.9 $\overline{[U, U]} = R$, hence $[[U, U], R] = [R, R] = [K, K] + [K, S]$. But $[[U, U], K + S] = [[U, U], K] + [[U, U], S]$; equating the skew and symmetric parts we see that $[[U, U], K] = [K, K]$. However, by the Jacobi identity, $[[U, U], K] \subset U$ leading to the contradiction $U = [K, K]$. Therefore we have proved $[[U, U],[U, U]] = (0)$.

The problem of examining the Lie simplicity of $[K, K]$ has been reduced to proving that if V is a Lie ideal of $[K, K]$ for which $[V, V] = (0)$ then $V = (0)$. Were this true, by Theorem 2.14 for any proper Lie ideal U of $[K, K]$, $[[U, U],[U, U]] = (0)$ hence we would get $[U, U] = (0)$ and so $U = (0)$. This explains our concern with Lie ideals V of $[K, K]$ such that $[V, V] = (0)$.

We prove a remark -- actually proved earlier-- namely:

Remark: If $a \in K$ satisfies $aKa = (0)$ then $a = 0$.

Proof. Let $s \in S$; then $sas \in K$ hence $asasa = 0$. Given $r \in R$, $r = s+k$, $s \in S$ and $k \in K$ whence $arara = a(s+k)a(s+k)a = 0$. By Lemma 1.1, since in aR every element has cube 0, $a = 0$.

LEMMA 2.10. Let V be a Lie ideal of $[K, K]$ such that $[V, V] = (0)$. Then V has no non-zero nilpotent elements.

Proof. Let $u \neq 0 \in V$. Then for any $k \in [K, K]$, $u(uk - ku) - (uk - ku)u = 0$, that is

$$(1) \qquad u^2k - 2uku + ku^2 = 0.$$

Thus we see that $u^2[K, K] \subset Ru$. If $k = ub - bu$, $b \in K$ then $u^2(ub - bu) \subset Ru$, hence $u^3K \subset Ru$. Finally, if $s \in S$, $us + su \in K$ so that $u^3(us + su) \subset Ru$ leading to $u^4S \subset Ru$. But then $u^4R \subset Ru$ so that $Ru^4R \subset Ru$; hence either $u^4 = 0$ or $Ru = R$. Similarly, if $u^4 \neq 0$ we get $uR = R$; therefore $u \in V$ implies $u^4 = 0$ or u is invertible.

Suppose that $u \in V$ is not invertible; then $u^4 = 0$ hence $u^2[K, K]u^3 \subset Ru\,u^3 = (0)$, so that $u^2[K, K]u^3 = (0)$. In particular, for $b \in K$ then $u^2(ub - bu)u^3 = 0$, that is $u^3Ku^3 = (0)$. Since $u^3 \in K$ we have seen this implies that $u^3 = 0$.

Let N be the set of nilpotent elements in V. Since $[V, V] = 0$, clearly N is closed under addition.

We claim that N is a Lie ideal of $[K, K]$. If the characteristic of R is 3, then for $a \in V$, $[a, [a, [a, k]]] = 0$ implies $a^3k = ka^3$ for all $k \in K$, hence $a^3 \in Z \cap K = (0)$ in which case $N = V$ so is a Lie ideal of $[K, K]$. If the characteristic of R is not 3, then if $u \in N$, since u commutes with $uk - ku$, for $k \in [K, K]$, $u^2k - ku^2 = 2u(uk - ku)$, $0 = u^3k - ku^3 = 3u^2(uk - ku)$. If $u^2 \neq 0$ then $uk - ku$ can not be invertible since it annihilates u^2, so is nilpotent. If $u^2 = 0$ then $u(uk - ku) = 0$ so that again $uk - ku$ is nilpotent. Thus N is a Lie ideal of $[K, K]$.

Let $u, v \in N$; then for $k \in [K,K]$, from (1), by multiplying by u from left and right, $u^2ku^2 = 0$, so $u^2(vb - bv)u^2 = 0$ for all $b \in K$. Thus $u^2vbu^2 = u^2bvu^2$, and so since u, v commute and $v^3 = 0$, $u^2v^2Ku^2v^2 = (0)$. Also, for $s \in S$, $v^2s - sv^2 \in [K,K]$ so that $u^2(v^2s - sv^2)u^2 = 0$, leading to $u^2v^2Su^2v^2 = (0)$. This gives $u^2v^2Ru^2v^2 = (0)$, hence $u^2v^2 = 0$. Linearizing on u, $(uw + wu)v^2 = 0$, that is $uwv^2 = 0$. Linearizing on v we get $uwvx = 0$ for $u, w, v, x \in N$. But this says $uvwN = (0)$, whence by Lemma 2.7, $uvw = 0$, hence $uv = 0$ hence $u = 0$. (Again using Lemma 2.7.) Thus $N = (0)$.

Note that in the course of proving the result we showed that if the characteristic of R is 3 then $[V,V] = (0)$ implies $V = (0)$ for V a Lie ideal of $[K,K]$. So in what follows <u>we may assume that R is of characteristic not 3</u>.

LEMMA 2.11. *If V is a Lie ideal of* $[K,K]$ *such that* $[V,V] = (0)$ *then* $V = (0)$.

Proof. In the proof of the previous lemma we got that the non-zero elements of V are invertible. Since for $u \in V$, $k \in K$, $uk - ku \in [K,K]$, $[u,[u,[u,k]]] = 0$. Writing $d(k) = uk - ku$, $d^3(k) = 0$ for $k \in K$.

Now $kuk \in K$, hence $d^3(kuk) = 0$. Working this out we get $3(d^2(k)ud(k) + d(k)ud^2(k)) = 0$, so applying d to this we get $d^2(k)ud^2(k) = 0$. Since $d^2(k) \in V$, either it is 0 or invertible. We thus get $d^2(k) = 0$ for all $k \in K$. Thus $d^2(kuk) = 0$, hence $d(k)ud(k) = 0$. If $k \in [K,K]$, then $d(k) \in V$ so is 0 or invertible. The net result is $d(k) = 0$ for $k \in [K,K]$. But then u centralizes $[K,K]$, hence $\overline{[K,K]} = R$, putting $u \in Z$, a contradiction. Thus $V = (0)$.

We have all the pieces to prove the basic

THEOREM 2.15. *If* R *is a simple ring of characteristic not 2 and if* $\dim R/Z > 16$ *then any proper Lie ideal of* $[K, K]$ *is in* Z.

Proof. In case the involution is of the second kind this is contained in Theorem 2.1. If the involution is of the first kind and if U is a proper Lie ideal of $[K, K]$, by Theorem 2.14, $[[U, U], [U, U]] = (0)$, hence by Lemma 2.11 $[U, U] = (0)$. By Lemma 2.11 again we get $U = (0)$.

Bibliography

1. W. Baxter, "Lie simplicity of a special class of associative rings II," *Trans. Amer. Math. Soc.* 87 (1958), pp. 63-75.

2. I. N. Herstein, "Lie and Jordan systems in simple rings with involution," *Amer. Journal Math.* 78 (1956), pp. 629-649.

CHAPTER 3

JORDAN HOMOMORPHISMS AND DERIVATIONS

Having studied the Jordan structure of certain rings and subsets of rings we now turn to the study of mappings related to the Jordan structure. We shall say that R is _2-torsion free_ if in R $2x = 0$ forces $x = 0$.

DEFINITION. A ring R is a _prime ring_ if $aRb = (0)$ implies that $a = 0$ or $b = 0$.

The definition just given is equivalent to:

1. The product of two non-zero 2-sided ideals of R is not (0).
2. The left annihilator of a non-zero left ideal is (0).

DEFINITION. A mapping ' from a ring R into a ring R' which is 2-torsion free is a _Jordan homomorphism_ if $(a+b)' = a'+b'$ and $(ab + ba)' = a'b' + b'a'$.

We shall define a Jordan homomorphism even when there is 2-torsion a little later. When R' is 2-torsion free note that we could have defined a Jordan homomorphism by merely insisting that $(a^2)' = (a')^2$.

We begin with

LEMMA 3.1. _If ' is a Jordan homomorphism of_ R _into_ R', _which is 2-torsion free then_ $(aba)' = a'b'a'$ _for all_ $a, b \in R$.

Proof. Let $x \circ y = xy + yx$; then $(a \circ (a \circ b))' = a' \circ (a' \circ b')$. Evaluating, using $(a^2)' = (a')^2$ we get $2(aba)' = 2a'b'a'$ hence $(aba)' = a'b'a'$.

We use the lemma to define Jordan homomorphism even in the case of 2-torsion by defining a Jordan homomorphism $'$ to be an additive mapping such that $(a^2)' = (a')^2$ and $(aba)' = a'b'a'$. In the 2-torsion free case this is the same as the original definition we gave.

Linearizing $(aba)' = a'b'a'$ on a we get that $(abc + cba)' = a'b'c' + c'b'a'$. We use this to prove

LEMMA 3.2. *If* $'$ *is a Jordan homomorphism of* R *into* R' *then* $((ab)' - a'b')((ba)' - a'b') = 0$ *and* $((ba)' - a'b')((ab)' - a'b') = 0$ *for all* $a, b \in R$.

Proof. We expand

$$((ab)' - a'b')((ab)' - b'a') = ((ab)')^2 + a'(b')^2a' - (ab)'b'a' - a'b'(ab)'$$
$$= ((ab)^2 + ab^2a - abba - (ab)^2)'$$
$$= 0 .$$

Similarly for the other assertion.

Note that this is enough to answer the question about the nature of Jordan homomorphisms for the special case in which R' has no zero divisors, for then we get $(ab)' = a'b'$ or $(ab)' = b'a'$ for every pair $a, b \in R$. From this it is an easy exercise to prove that $(xy)' = x'y'$ for all $x, y \in R$ or $(xy)' = y'x'$ for all $x, y \in R$, that is, the mapping is either a homomorphism or an anti-homomorphism.

We want to show that this conclusion holds in the much more general case wherein R' is any prime ring, provided the mapping is *onto*.

DEFINITION. For $a, b \in R$, $a^b = (ab)' - a'b'$ and $a_b = (ab)' - b'a'$.

A few easy properties of these symbols:

1. $a^b = -b^a$ (this is merely a restatement of the fact that $'$ is a Jordan homomorphism). Similarly $a_b = -b_a$.

2. $[a', b'] = a_b - a^b$

3. $[a, b]' = a_b + a^b$

4. $a^b a_b = a_b a^b = 0$ (Lemma 3.2)

LEMMA 3.3. *For* $a, b, r \in R$,

$$([a, b]r)' = r'a^b + a_b r'$$

and

$$(r[a, b])' = a^b r' + r'a_b .$$

Proof. $r'a^b + a_b r' = r'(ab)' - r'a'b' - b'a'r' + (ab)'r'$ $= (rab + abr)' - (rab + bar)' = ((ab - ba)r)'$. Similarly for the other half of the lemma.

LEMMA 3.4. *If* $'$ *is a Jordan homomorphism of* R *into* R' *then* $a_b r'a^b + a^b r'a_b = 0$.

Proof. From Lemma 3.3,

1. $$r'a^b + a_b r' = ([a, b]r)'.$$

Multiply by a^b from the left, using $a^b a_b = 0$ we get

2. $$a^b r'a^b = a^b([a, b]r)'.$$

Multiplying (1) from the right by a_b we get

3. $$a_b r'a_b = ([a, b]r)'a_b.$$

In $a^b r' + r'a_b = (r[a, b])'$ replace r by $[a, b]r$; using 1, 3 and Lemma 3.1 we get

$$a^b r'a^b + a_b r'a_b = [a, b]'r'[a, b]' = (a^b + a_b)r'(a^b + a_b).$$

Expanding out we end up with $a^b r'a_b + a_b r'a^b = 0$, the assertion of the lemma.

We now prove the

THEOREM 3.1. Let ' be a Jordan homomorphism of R onto R'. If R' is a prime ring then ' is either a homomorphism or an anti-homomorphism.

Proof. By the previous lemma

(1) $a^b r' a_b + a_b r' a_b = 0.$

Suppose that $(c^d)^2 \neq 0$ for some $c, d \in R$. Multiplying (1) by c^d (where $a = c$, $b = d$ in (1)), we get $(c^d)^2 R' c_d = 0$. Since R' is prime and $(c^d)^2 \neq 0$, then $c_d = 0$.

Thus if $(c^d)^2 \neq 0$ we can conclude that $c_d = 0$.

We linearize (1) on b to get

(2) $a^b r' a_x + a^x r' a_b + a_b r' a^x + a_x r' a^b = 0.$

and linearize $a^b a_b = 0$ to get

(3) $a^b a_x + a^x a_b = 0.$

Since $(c^d)^2 \neq 0$, $c_d = 0$ we get on putting $a = c$, $b = d$ in (2) and (3)

(4) $c^d r' c_x + c_x r' c^d = 0$

$c^d c_x = 0$

for all $r, x \in R$.

Multiplying the first equation of (4) from the left by c^d, we get since $c^d c_x = 0$, $(c^d)^2 R' c_x = 0$; since R' is prime and $(c^d)^2 \neq 0$, $c_x = 0$.

Thus, if $(c^d)^2 \neq 0$ then $c_x = 0$ for all $x \in R$. Also $(d^c)^2 = (-c^d)^2 \neq 0$ hence $d_x = 0$ for all $x \in R$. Similarly if $(c_d)^2 \neq 0$ then $c^x = 0$ for all $x \in R$.

We linearize (2) on a to get

$$(5) \qquad a^b r' y_x + a^x r' y_b + y^b r' a_x + y^x r' a_b$$
$$+ a_b r' y^x + a_x r' y^b + y_b r' a^x + y_x r' a^b = 0.$$

If $(c^d)^2 \neq 0$ then $c_x = d_x = 0$ for all x. Putting $a = c$, $b = d$ in (5) yields

$$(6) \qquad c^d r' y_x + y_x r' c^d = 0$$

for all $x, y \in R$.

Now from $x^y x_y = 0$ linearized on x, y we get

$$(7) \qquad x^y w_z + w^y x_z + x^z w_y + w^z x_y = 0.$$

In (7) putting $w = c$, $z = d$ we get since $c_r = d_s = 0$ for all r, s, $c^d x_y = 0$. Multiply (6) by c^d to now get $(c^d)^2 R' y_x = (0)$ hence $y_x = 0$ for all $x, y \in R$. That is, $(yx)' = x'y'$, which is to say that ' is an anti-homomorphism. Similarly, if for some c, d, $(c_d)^2 \neq 0$ then ' must be a homomorphism.

We are left to consider the situation in which $(x^y)^2 = 0$, $(x_y)^2 = 0$ for all $x, y \in R$. But then, since $[x', y'] = x_y - x^y$, and since $x_y x^y = x^y x_y = (x^y)^2 = (x_y)^2 = 0$, $[x', y']^2 = 0$ for all $x'y' \in R'$. Since all commutators in R' have square 0, if $(a')^2 = 0$, we get $a'R'$ is a nil ideal of index 3; by Lemma 1.1 R' would have a nilpotent ideal which is impossible in a prime ring. Thus R' has no nilpotent elements, hence $[x', y'] = 0$, that is R' is commutative. But being prime it would have no zero-divisors, so as remarked at the end of the proof of Lemma 3.2 this forces ' to be a homomorphism. We have now proved the theorem.

We consider mappings now that are additive and preserve n^{th} powers.

DEFINITION. A mapping φ of R into R' is an <u>n-Jordan mapping</u> if φ is additive and $\varphi(a^n) = \varphi(a)^n$ for all $a \in R$.

THEOREM 3.2. *Let φ be an n-Jordan mapping from R onto a prime ring R' whose characteristic is 0 or larger than n. Then if R has a unit element, $\varphi = \epsilon\tau$ where τ is either a homomorphism or an anti-homomorphism and ϵ is an (n-1)-st root of unity lying in the center of R'.*

Proof. Since $\varphi(x^n) = \varphi(x)^n$ linearizing on x we get

$$(1) \qquad \sum_{\pi \in S_n} \varphi(x_{\pi(1)} \cdots x_{\pi(n)}) = \sum_{\pi \in S_n} \varphi(x_{\pi(1)}) \cdots \varphi(x_{\pi(n)}),$$

where S_n is the symmetric group of degree n.

Let $\alpha = \varphi(1)$; thus $\alpha^n = \varphi(1)^n = \varphi(1) = \alpha$. In (1) put $x_1 = x$, $x_2 = \ldots x_n = 1$. We get

$$(2) \qquad n!\,\varphi(x) = (n-1)!\,(\alpha^{n-1}\varphi(x) + \alpha^{n-2}\varphi(x)\alpha + \ldots + \varphi(x)\alpha^{n-1}).$$

Hence $n\varphi(x) = (\alpha^{n-1}\varphi(x) + \ldots + \varphi(x)\alpha^{n-1})$, since the characteristic of R' exceeds n. Commuting this with α, since $\alpha^n = \alpha$ we get $(n-1)(\alpha\varphi(x) - \varphi(x)\alpha) = 0$, hence $\alpha\varphi(x) = \varphi(x)\alpha$. Therefore α is an (n-1)-st root of unity and lies in the center of R'.

We return to (1) putting $x_1 = x$, $x_2 = y$ and $x_3 = \ldots = x_n = 1$. (1) then gives, from the condition on the characteristic, $\varphi(xy + yx) = \alpha^{n-2}(\varphi(x)) + \varphi(y)\varphi(x))$. Let $\tau = \alpha^{n-1}\varphi$; then τ is a Jordan homomorphism of R onto R'. By Theorem 3.1 it is either a homomorphism or anti-homomorphism. This completes the proof of the theorem.

We now turn to the question of Jordan derivations. We assume all rings to be 2-torsion free.

DEFINITION. A <u>Jordan derivation</u> $'$ on R is an additive mapping of R into itself such that $(a^2)' = aa' + a'a$.

Note that this is equivalent to

$$(a \circ b)' = a' \circ b + a \circ b' .$$

We begin with

LEMMA 3.5. *If $'$ is a Jordan derivation on R then $(aba)' = a'ba + ab'a + aba'$ for all $a, b \in R$.*

Proof. We compute $(a \circ (a \circ b))'$ in two ways.

$$(a^2 \circ b + 2aba)' = (a \circ (a \circ b))' = a' \circ (a \circ b) + a \circ (a \circ b)'.$$

Evaluating we get the result.

Linearizing we have that

COROLLARY. *For $a, b, c \in R$*

$$(abc + cba)' = a'bc + ab'c + abc' + c'ba + cb'a + cba'.$$

Essential in what follows is

LEMMA 3.6. For $a, b \in R$,

$$((ab)' - a'b - ab')(ab - ba) = 0.$$

Proof. Let $w = ((ab)ab + ba(ab))' = (ab)'ab + (ab)a'b + abab' + b'a(ab) + ba'ab + ba(ab)'$. But $w = (ab)^2 + (ba^2b)' = ab(ab)' + (ab)'ab + b'a^2b + b(aa' + a'a)b + ba^2b'$. Comparing the results we get the lemma.

DEFINITION. Let $a^b = (ab)' - a'b - ab'$.

In these terms the lemma reads $a^b(ab - ba) = 0$. Note that a^b is linear in a and b and $a^b = -b^a$.

Linearizing Lemma 3.6 on a we get the

COROLLARY 1. $a^b(cb - bc) + c^b(ab - ba) = 0$ <u>for</u> $a, b, c \in R$.

COROLLARY 2. $a^c(ab - ba) + a^b(ac - ca) = 0$.

This last corollary is merely linearization on b.

LEMMA 3.7. <u>If</u> R <u>is a prime ring and if</u> $ab = 0$ <u>then</u> $a^b = 0$.

<u>Proof</u>. If $ab = ba = 0$ then by the corollary to Lemma 3.6, $a^b(cb - bc) = 0$ for all c. b is not in Z since Z has no zero divisors; but then $a^b(cb - bc) = 0$ leads us to $a^b(cxb - bcx) = 0$, hence $a^bR(xb - bx) = (0)$. This gives $a^b = 0$.

Suppose $ab = 0$, $ba \neq 0$. Then by the corollary 2 to Lemma 3.6, $a^b(ac - ca)b = 0$, hence $a^b aRb = 0$. Since R is prime, $a^b a = 0$ for any b with $ab = 0$. But then since $a^b = (ab)' - a'b - ab'$, we get $(a'b + ab')a = 0$. For any $r \in R$, $abrb = 0$, hence $(a'brb + a(brb)')a = 0$; since $(brb)' = b'rb + br'b + brb'$, we get $(a'brb + ab'rb)a = 0$, whence $(a'b + ab')Rba = 0$ and since $ba \neq 0$ it follows that $a'b + ab' = 0$. Thus $a^b = 0$.

Note that in this case $b^a = -a^b = 0$, so that $(ba)' = b'a + ba'$.

LEMMA 3.8. If $ab = 0$ then for $c \in R$, R prime, $(c(ba))' = c'(ba) + c(ba)'$.

<u>Proof.</u> By the corollary to Lemma 3.5

$$(cba)' = (cba + abc) = c'ba + cb'a + cba' + a'bc + ab'c + abc'$$

$$= c'(ba) + c(b'a + ba') + (a'b + ab')c$$

By Lemma 3.7, $a'b + ab' = 0$ and $b'a + ba' = (ba)'$. Thus the lemma.

DEFINITION. $V = \{x \in R \mid (ax)' = a'x + ax' \text{ all } a\}$.

Note that by Lemma 3.8, if $ab = 0$ then $ba \in V$.

We need one more lemma.

LEMMA 3.9. *If* $v, w \in V$ *then* $a^b(vw - wv) = 0$ *for all* $a, b \in R$.

Proof. By the corollary to Lemma 3.5

$$a^b(bv - vb) + v^b(ab - ba) = 0,$$

hence, since $v^b = 0$, $a^b(bv - vb) = 0$ for all b. We linearize this on b to get $a^b(cv - vc) + a^c(bv - vb) = 0$. Putting $c = w \in V$, since $a^w = 0$ we get $a^b(wv - vw) = 0$.

We now prove the

THEOREM 3.3. *If* R *is a prime ring of characteristic different from* 2 *then any Jordan derivation of* R *is a derivation of* R.

Proof. Suppose that $a^b \neq 0$. Given $c \in R$, since $c^b(cb - bc) = 0$ (Lemma 3.6), we get, for any $r \in R$, $v = (cb - bc)rc^b \in V$ by Lemma 3.8. Similarly for $s \in R$, $(ab - ba)sa^b \in V$. Using these v, w in Lemma 3.9 in conjunction with Lemma 3.6 we get $a^b(cb - bc)Rc^b(ab - ba)Ra^b = (0)$. Since R is prime and $a^b \neq 0$ we get $a^b(cb - bc) = 0$ or $c^b(ab - ba) = 0$; but $a^b(cb - bc) + c^b(ab - ba) = 0$ by the corollary to Lemma 3.6, whence $a^b(cb - bc) = 0$ for all $a, b, c \in R$.

If $b \notin Z$ since the annihilator of all $cb - bc$, $c \in R$ is an ideal we get $a^b = 0$. If $Z \neq R$ and $b \in Z$, choosing $c \notin Z$ then $b + c \notin Z$ hence $0 = a^c = a^{b+c} = a^b + a^c$ leading to $a^b = 0$. Finally if $Z = R$, then the definition of a Jordan derivation yields $2(ab)' = 2(a'b + ab')$, whence by our assumption on the characteristic we conclude that $(ab)' = a'b + ba'$.

Thus in all cases $a^b = 0$ for all $a, b \in R$, therefore $'$ is a derivation of R.

If one checks the proof given one sees that the assumption that the characteristic of R be different from 2 enters only at two points: in proving $(aba)' = a'ba + ab'a + aba'$ and at the very end of the argument just given.

If we redefine a Jordan mapping by $(a^2)' = aa' + a'a$ and $(aba)' = a'ba + ab'a + aba'$, then in characteristic not 2 we have imposed no extra restriction yet in characteristic 2 it allows us to conclude

THEOREM 3.4. _If_ R _is a prime ring and_ $'$ _is a Jordan derivation (as newly redefined) of_ R _then_ $'$ _is a derivation except if_ R _is both commutative (and so an integral domain) and of characteristic_ 2.

At this point we should like to redo the theorem on Jordan homomorphisms from a slightly different point of view. It follows fairly easily from a simple result about prime rings but suffers from the defect that we must insist that the characteristic be different from 2. Outside of this lemma which we are about to prove we only need the result of Lemma 3.1.

LEMMA 3.10. _Let_ R _be a prime ring of characteristic not_ 2 _and suppose_ $a, b \in R$ _are such that_ $arb + bra = 0$ _for all_ $r \in R$. _Then either_ $a = 0$ or $b = 0$.

Proof. By assumption $arb + bra = 0$. Replace r by sat with $s, t \in R$. We get $asatb + bsata = 0$. But $bsa = -asb$ and $atb = -bta$; substituting these we get $-asbta - asbta = 0$, that is, $2aRbRa = (0)$. Since R is prime of characteristic not 2 this immediately gives $a = 0$ or $b = 0$.

We now reprove Theorem 3.1, assuming that $'$ is a Jordan homomorphism of R onto R', R' of characteristic not 2.

Consider $u = (ab)' - a'b'$, $v = (ab)' - b'a'$; then

$$\begin{aligned} ur'v &= ((ab)' - a'b')r'((ab)' - b'a') \\ &= (ab)'r'(ab)' + a'b'r'b'a' - a'b'r'(ab)' - (ab)'r'b'a' \\ &= (abrab)' + (abrba)' - a'b'r'(ab)' - (ab)'r'b'a' , \end{aligned}$$

using Lemma 3.1. Similarly,

$$vru = (abrab)' + (barab)' - (b'a'r'(ab)' - (ab)'r'a'b'.$$

Thus

$$\begin{aligned} urv + vru = &\{2abrab + abrba + barab\}' \\ &- \{(a'b' + b'a')r'(ab)' + (ab)'r'(a'b' + b'a')\} \\ = &\{2abrab + abrba + barab\}' \\ &- \{(ab + ba)r(ab) + (ab)r(ab + ba)\} \\ = &\{2abrab + abrba + barab\}' \\ &- \{(ab + ba)r(ab) + (ab)r(ab + ba)\}' \end{aligned}$$

(using the corollary to Lemma 3.1)

$$= 0.$$

Since $ur'v + vr'u = 0$ for all $r' \in R'$, using Lemma 3.10 we have $u = 0$ or $v = 0$, that is, $(ab)' = a'b'$ or $(ab)' = b'a'$.

Given $a \in R$ let $W_a = \{b \in R \mid (ab)' = a'b'\}$ and $V_a = \{b \in R \mid (ab)' = b'a'\}$. Clearly each of W_a and V_a are additive subgroups of R. Moreover, by the discussion given, R is the set-theoretic union of W_a and V_a. But a group can not be the set-theoretic union of two proper subgroups, hence $W_a = R$ or $V_a = R$, that is, $(ab)' = a'b'$ for all $b \in R$ or $(ab)' = b'a'$ for all $b \in R$. Let $W = \{a \in R \mid (ax)' = a'x' \text{ for all } x \in R\}$ and $V = \{a \in R \mid (ax)' = x'a' \text{ for all } x \in R\}$; both W and V are additive subgroups of R and, from the above, their union is R. In consequence, $W = R$ or $V = R$. If $W = R$ then $(xy)' = x'y'$ for all $x, y \in R$ and $'$ is a homomorphism; if $V = R$ then $(xy)' = y'x'$ for all $x, y \in R$ and $'$ is an anti-homomorphism. This is Theorem 3.1.

We close the chapter with a discussion, without proofs, of some highly interesting theorems proved in the last few years by Martindale.

The first of these concerns Jordan maps of the symmetric elements of a ring with involution, that is, additive maps ϕ defined on the symmetric elements such that $\phi(s^2) = \phi(s)^2$ for s symmetric. He proves:

Let R be a ring with involution, $*$, such that:

a) R contains three non-zero orthogonal idempotents e_1, e_2, e_3 such that $e_1 + e_2 + e_3 = 1$,

b) $e_iRe_jRe_i = e_iRe_i$ for $i, j = 1, 2, 3$.

Let ϕ be a Jordan homomorphism of the symmetric elements S of R in a ring R' whose characteristic is not 2. Then ϕ can be extended uniquely to an associative homomorphism Φ of R into R'.

Martindale has obtained very decisive results about Lie homomorphisms of one ring into another. A Lie homomorphism ϕ is an additive map such that $\phi(ab - ba) = \phi(a)\phi(b) - \phi(b)\phi(a)$. We now give the statement of some of these.

1. Let ϕ be a Lie isomorphism of a primitive ring R onto a primitive ring R', where R is not of characteristic 2 or 3. Suppose that R contains three orthogonal idempotents whose sum is 1. Then ϕ is of the form $\sigma + \tau$ where σ is either an isomorphism or the negative of an anti-isomorphism of R into a primitive ring L' containing R' and τ is an additive mapping of R into the center of L' taking all commutators $ab - ba$ into 0.

In an unpublished result (which will appear) he has extended this result significantly in the following way:

1. R and R' are merely assumed to be prime rings.

2. Instead of three orthogonal idempotents he only requires that $1 \in R$ and R have an idempotent $e \neq 0, 1$.

With these weakened hypotheses he still gets the same conclusion about the nature of Lie isomorphisms.

In the case when R and R' are simple and R is of characteristic not 2 or 3 with unit element and having an idempotent $e \neq 0, 1$ a little more can be said, namely, that a Lie isomorphism of R onto R' is of the form $\sigma + \tau$, σ an isomorphism or the negative of an anti-isomorphism of R onto R' (here one does not need the extension L' or R') and τ is an additive map of R into the center of R' vanishing on commutators ab - ba.

A student of Martindale, R. A. Howland has recently proved a related result: Let R be a simple ring of characteristic not 2 or 3 with three non-zero orthogonal idempotents whose sum is 1 and let T be a simple ring with 1. Then any Lie isomorphism of [R, R] onto [T, T] can be uniquely extended to an associative isomorphism of R onto T.

These results give much information about the automorphisms of R and [R, R] as Lie algebras in terms of associative automorphisms and trace-like mappings of R itself.

Finally there is the closely connected question of Lie derivations, that is, additive mappings, δ, such that

$$\delta([a, b]) = [\delta(a), b] + [a, \delta(b)].$$

Martindale has proved:

Let R be a primitive ring containing an idempotent $e \neq 0, 1$ and suppose that the characteristic of R is not 2. Then any Lie derivation of R is of the form D + T where D is an associative derivation of R into a primitive ring $\overline{R} \supset R$ and T is an additive mapping of R into the center of $\overline{R}$ such that $T(ab - ba) = 0$ for all $a, b \in R$.

Bibliography

1. I.N. Herstein, "Jordan homomorphisms", Trans. Amer. Math. Soc. 81 (1956), pp. 331-351.

2. I.N. Herstein, "Jordan derivations of prime rings", Proc. Amer. Math. Soc. 8 (1957), pp. 1104-1110.

3. I.N. Herstein, "Special Jordan maps", to appear in Anais Acad. Brasil.

4. W. Martindale, "Jordan homomorphisms of the symmetric elements of a ring with involution", Journal of Algebra 5 (1967), pp. 232-249.

5. W. Martindale, "Lie isomorphisms of primitive rings", Proc. Amer. Math. Soc. 14 (1963), pp. 909-916.

6. W. Martindale, "Lie isomorphisms of simple rings", to appear in Jour. London Math. Soc.

7. W. Martindale, "Lie derivations of primitive rings", Michigan Math. Journal 11(1964), pp. 183-187.

8. M.F. Smiley, "Jordan homomorphisms onto prime rings", Proc. Amer. Math. Soc. 8 (1957), pp. 426-429.

CHAPTER 4

GOLDIE'S THEOREMS

In this chapter we shall make a study of rings satisfying certain ascending chain conditions. In the non-commutative case -- and this is really the only case with which we shall be concerned -- the decisive and incisive results are three theorems due to Goldie. The main part of the chapter will be taken up with a presentation of these.

DEFINITION. An element a in the ring R is <u>regular</u> if it is neither a left nor right zero divisor in R.

DEFINITION. A ring $Q(R) \supset R$ is said to be <u>left quotient ring</u> for R if:

1. $a \in R$ is regular then it is invertible in $Q(R)$
2. every $x \in Q(R)$ is of the form $a^{-1}b$, where $a, b \in R$ and a is regular.

We begin the discussion with a theorem due to Ore.

THEOREM 4.1. <u>A necessary and sufficient condition that R have a left-quotient ring is: given $a, b \in R$, b regular, then there exists a regular element b_1 in R so that $b_1 a = a_1 b$ for some a_1 in R.</u>

<u>Proof</u>. If $Q(R)$ exists then for b regular in R, $ab^{-1} \in Q(R)$ so $ab^{-1} = b_1^{-1} a_1$ with b_1 regular in R. Thus $b_1 a = a_1 b$ as required.

Conversely, suppose the condition stated in the theorem holds in R. Let $\mathfrak{M} = \{(a,b) \mid a,b \in R,\ b \text{ regular}\}$. We define

$(a, b) \sim (c, d)$ if $d_1 a = b_1 c$ where $b_1 d = d_1 b$, b_1 regular. (And so d_1 is regular.) We claim this is independent of b_1, d_1 for if $b_2 d = d_2 b$ pick e_1, e_2 regular so that $e_2 b_2 = e_1 b_1$. Then

$$\begin{array}{c} e_1 b_1 d = e_1 d_1 b \\ \parallel \\ e_2 b_2 d = e_2 d_2 b . \end{array}$$

By the regularity of b, $e_1 d_1 = e_2 d_2$, hence from $d_1 a = b_1 c$, $e_2 d_2 a = e_1 d_1 a = e_1 b_1 c = e_2 b_2 c$ and since e_2 is regular, $d_2 a = b_2 c$.

We claim the relation defined on $\mathfrak{M}$ is an equivalence relation. Let the class of (a, b) be written as a/b.

If $b_1 d = d_1 b$ define $a/b + c/d = (d_1 a + b_1 c)/(b_1 d)$ and $(a/b)(c/d) = (a_1 c)/(d_1 b)$, where $d_1 a = a_1 d$.

One can easily verify that the equivalence classes form a ring with unit, $Q(R)$, which is a left quotient ring for R.

DEFINITION. If $S \subset R$ is a subset then $\ell(S) = \{x \in R \mid xS = (0)\}$, the <u>left-annihilator</u> of S and $r(S) = \{x \in R \mid Sx = (0)\}$, the <u>right annihilator</u> of S.

Clearly $\ell(S)$ is a left ideal of R, and $r(S)$ is a right ideal of R. We say that a left ideal λ of R is a <u>left annihilator</u> if $\lambda = \ell(S)$ for some $S \subset R$.

We now introduce the Goldie chain conditions.

DEFINITION. R satisfies the <u>Goldie chain conditions</u> if:

1. any ascending chain of left annihilators becomes stationary after some point.

2. R has no infinite direct sum of left ideals.

A ring satisfying the Goldie conditions will be called a <u>Goldie ring</u>.

LEMMA 4.1. Let R be a Goldie ring without zero divisors. Then R has a left quotient ring which is a division ring.

Proof. We shall show that the intersection of any two non-zero left ideals of R is not (0). Thus if, $a, b \in R$, neither 0, then $Ra \cap Rb \neq (0)$ so by Theorem 4.1 R will have a ring of left quotients which will obviously be a division ring.

We claim that there is a left ideal $U \neq (0)$ of R such that any two non-zero left ideals of R lying in U have a non-zero intersection. If not, there are two non-zero left ideals U_1, U_2 with intersection (0), $U_1 \cap U_2 = (0)$. If in U_1 there are two non-zero left ideals of R $U_1^{(1)}, U_1^{(2)}$ with intersection (0) then $U_1^{(1)} + U_1^{(2)} + U_2$ is a direct sum. Continue this on $U_1^{(1)}$, and so on, to get longer and larger direct sums. Since there are no infinite direct sums of left ideals of R we finally must stop. We thus produce the required left ideal U of R.

Suppose that λ, μ are non-zero left ideals of R. Pick $a \neq 0$ in U; then λa and μa are non-zero left ideals of R lying in U hence must have a non-zero intersection, so there is an $0 \neq x = \alpha a = \beta a$ where $\alpha \in \lambda$, $\beta \in \mu$. Since R has no zero divisors $0 \neq \alpha = \beta \in \lambda \cap \mu$, whence $\lambda \cap \mu \neq (0)$.

DEFINITION. A ring R will be called an Ore domain if it has no zero divisors and the intersection of any two non-zero left ideals of R is not zero.

Properly speaking such a ring should be called a left Ore domain but for the sake of brevity we drop the "left." Note that by Theorem 4.1 a ring is an Ore domain if it has a ring of left quotients which is a division ring.

Let R be an Ore domain and let D be the division ring of left quotients of R. If $k_1, \ldots, k_n \in D$ then $k_i = a_i^{-1} b_i$ with

$a_i, b_i \in R$. We claim that we can find an element $a \in R$ and $c_i \in R$ such that each $k_i = a^{-1}c_i$. We proceed by induction on n; suppose that t has been found in R such that $k_1 = t^{-1}d_1, \ldots, k_{n-1} = t^{-1}d_{n-1}$, where $d_1, \ldots, d_{n-1} \in R$. Now $k_n = a^{-1}b_n$; moreover there is an element $a \neq 0$ such that $a = st = ua_n$. Then $k_1 = a^{-1}sd_1, \ldots, k_{n-1} = a^{-1}sd_{n-1}$, $k_n = a^{-1}ub_n$ is the desired representation. We shall use this remark often so we record it as

Sublemma A. *If R is an Ore domain with a ring of left quotients D then, given $k_1, \ldots, k_n$ in D, we can find an element $a \in R$, and $c_1, \ldots, c_n \in R$ such that $k_i = a^{-1}c_i$ for $i = 1, 2, \ldots, n$.*

We now prove a condition that a ring be an Ore domain.

LEMMA 4.2. *Let R be a ring without zero divisors. Then the following are equivalent:*

1. *R is an Ore domain.*
2. *In R there is no infinite direct sum of left ideals.*
3. *If $\lambda \neq (0)$ is a fixed given left ideal of R then there is no infinite direct sum of left ideals of λ.*

Proof. That (1) implies (2) is trivial for any two left ideals in an Ore domain intersect.

(2) implies (3) for if $\lambda \neq (0)$ is a left ideal of R if $\lambda_1 \neq (0)$ is a left ideal of λ then $(0) \neq \lambda\lambda_1 \subset \lambda_1$ and $\lambda\lambda_1$ is a left ideal of R. So if $\lambda_1 + \ldots + \lambda_n$ is a direct sum of left ideals of λ then $\lambda\lambda_1 + \ldots + \lambda\lambda_n$ is one of left ideals of R.

(3) implies (1) for suppose $\lambda \neq (0)$ is a left ideal of R having no infinite direct sums. As in Theorem 4.1 we can find a left ideal $\lambda_1 \neq (0)$ of λ in which any two left ideals intersect. Since $\lambda\lambda_1 \subset \lambda_1$ we may assume that λ_1 is a left ideal of R. Let a, b be non-zero in R; pick $\alpha \neq 0$ in λ_1. Then $Ra\alpha, Rb\alpha$ are left ideals in λ_1 so intersect, that is there are $x, y \in R$ with $xa\alpha = yb\alpha \neq 0$. Since R is without zero divisors $xa = yb \neq 0$ follows.

We record one more fact about the relationship between an Ore domain and its ring of quotients.

LEMMA 4.3. _Let_ E _be an Ore domain,_ K _its ring of left quotients and suppose that_ $A \neq (0)$ _is a left ideal of_ E. _Then_ A _has_ K _as its ring of left quotients._

Proof. Given $k \in K$, $k = a^{-1}b$ where $a, b \in E$. Pick $c \neq 0$ in A; therefore there are elements $x, y \in R$, not zero, such that $xa = yc \neq 0$; also there are elements $r, s \in R$ such that $rxb = sc \neq 0$. Thus $k = a^{-1}b = (rxa)^{-1}rxb = (ryc)^{-1}sc = u^{-1}v$ where $u = ryc \in A$ and $v = sc \in A$. This proves the lemma.

We recall from Chapter 3 that a ring R is a prime ring if $aRb = (0)$ implies that $a = 0$ or $b = 0$. If $\lambda \neq (0)$ is a left ideal of R and $\rho \neq (0)$ is a right ideal of R where R is prime then $(0) \neq \rho\lambda \subset \lambda \cap \rho$, that is, every non-zero left ideal intersects every non-zero right ideal non-trivially.

In what follows R will be a prime Goldie ring. Let F be a maximal left annihilator in R and let $F_r = r(F)$. We shall maintain this notation. If $F_r = (0)$ we know from Lemma 4.1 that R has a ring of left quotients which is a division ring. So we suppose henceforth that $F_r \neq (0)$, $F \neq (0)$. We follow the proof given by Procesi.

DEFINITION. For $u \in F_r$ let $\lambda_u : F_r \to F_r$ defined by $\lambda_u(x) = ux$.

Let $E = \{\lambda_u \mid u \in F_r\}$; clearly E is a ring. However much more can be said.

THEOREM 4.2. E _is an Ore domain, and so has a ring of left quotients_ K _which is a division ring._

Proof. Since R is prime and $F_r \neq (0)$, $F_r^2 \neq (0)$, that is, $EF_r \neq (0)$; thus $E \neq (0)$. Let $\lambda_u \neq 0$, $\lambda_v \neq 0 \in E$ and suppose that $\lambda_u \lambda_v = 0$; then $uvF_r = (0)$. Since $\lambda_v \neq 0$, $vF_r \neq (0)$; but then since $R \neq \ell(vF_r) \supset F$ by the maximality of F we conclude that $\ell(vF_r) = F$. Since $u \in \ell(vF_r)$, u must be in F so $\lambda_u = 0$ contrary to assumption. We have shown E to be free of zero divisors.

Now claim that E has no infinite direct sum of left ideals for if $E\lambda_{u_1} + \ldots + E\lambda_{u_k}$ is a direct sum, since $\lambda_{u_i} \neq 0$, $F_r u_i \approx E\lambda_{u_i}$ from the definition of the λ's. Therefore $F_r u_1 + \ldots + F_r u_k$ is a direct sum. We claim that $Ru_1 + \ldots + Ru_k$ is a direct sum of left ideals of R for if $r_1 u_1 + \ldots + r_k u_k = 0$ with $r_1 u_1 \neq 0$, from the primeness of R, since $F_r r_1 u_1 \neq (0)$ there is a $v \in F_r$ with $vr_1 u_1 \neq 0$. But then $vr_1 u_1 + \ldots + vr_k u_k = 0$ with $vr_1 u_1 \neq 0$ contradicting the directness of $F_r u_1 + \ldots + F_r u_k$. Since R has no infinite direct sum of left ideals we see that E also has no infinite direct sums of left ideals. By Lemma 4.2 E is an Ore domain. By Theorem 4.1, E has a ring of left quotients which is a division ring K.

We return to our study of prime rings, using the symbols E, K, F_r, etc. as they were used prior to and in Theorem 4.2. We first make the

DEFINITION. $V = K \otimes_E F_r$.

Notice that V is a left vector space over the division ring K. Since F_r is a right ideal of R, V is a right R-module where $(k \otimes a)s = k \otimes (as)$ for $a \in F_r$, $s \in R$. Since R is prime, V is a faithful R-module. The elements of R induce K-linear transformations on V so we can consider R to be contained in $L(V)$, the ring of K-linear transformations on V.

Let $j: F_r \to K \otimes F_r = V$ defined by $j(u) = 1 \otimes u$ for $u \in F_r$; j is an injection of F_r into V. Given $v \in V$, $v = \Sigma k_i \otimes u_i$ and since, by Sublemma A, $k_i = \lambda_a^{-1} \lambda_{b_i}$, $a, b_i \in F_r$, we get

$$v = \lambda_a^{-1} \Sigma \lambda_{b_i} \otimes u_i = \lambda_a^{-1} \Sigma (\otimes \lambda_{b_i} u_i = \lambda_a^{-1}(1 \otimes \Sigma b_i u_i)),$$

that is, every $v \in V$ has the form $v = \lambda_a^{-1} j(w)$ where $a, w \in F_r$.

We now establish some vital facts about V which will allow us to get the first Goldie theorem.

THEOREM 4.3. V <u>is finite-dimensional over</u> K.

<u>Proof</u>. As we have just seen, $V = Kj(E_r)$ so we can find a basis of V over K of the form $j(u_\alpha)$, $u_\alpha \in F_r$. If $j(u_1), \ldots, j(u_n)$, $u_i \in F_r$, are linearly independent over K, we claim that $Ru_1 + \ldots + Ru_n$ is a direct sum. For suppose that $a_1 u_1 + \ldots + a_n u_n = 0$; if $a_1 u_1 \neq 0$, say, then we can find an element $b \in F_r$ such that $ba_1 u_1 \neq 0$, by the primeness of R. Thus $ba_1 u_1 + \ldots + ba_n u_n = 0$; since $ba_i \in F_r$ this says that $\lambda_{ba_1} j(u_1) + \ldots + \lambda_{ba_n} j(u_n) = 0$ contradicting the linear independence of the $j(u)$'s over K. Therefore the $a_i u_i = 0$ and the sum is direct. Since R has no infinite direct sums of left-ideals we conclude that V is finite-dimensional over K.

Since V is n-dimensional, say, over K and R acts faithfully on V as a ring of linear transformations we get that $R \subset K_n$, the ring of all $n \times n$ matrices over K. This is not at all obvious <u>ab initio</u>, and gives enough information to prove most of what we would usually want. However, we want to push deeper and to find out exactly how R sits in K_n.

We prove an easy consequence of Theorem 4.3.

LEMMA 4.4. *If* $j(u_1), \ldots, j(u_n) \in V$ $(u_i \in F_r)$ *is a basis of* V *over* K *then* $Ru_1 + \ldots + Ru_n$ *intersects every non-zero left-ideal of* R *in a non-zero fashion.*

Proof. Let $I = Ru_1 + \ldots + Ru_n$ and suppose that $J \neq (0)$ is a left-ideal of R such that $I \cap J = (0)$. Since R is prime $J \cap F_r \neq (0)$; let $u \neq 0$ be in $J \cap F_r$.

We claim that $j(u_1), \ldots, j(u_n)$, $j(u)$ are linearly independent over K; this has already been proved in the course of proving Theorem 4.3. Since this contradicts the fact that $j(u_1), \ldots, j(u_n)$ is a basis of V we get the lemma.

DEFINITION. We say that a left-ideal I of R *meets* all left-ideals of R if, given $J \neq (0)$ a left-ideal of R then $I \cap J \neq (0)$.

Sublemma B. *Let* V *be a vector space of dimension* n *over a division ring* K *and suppose that* $T_1, \ldots, T_k$ *are linear transformations on* V *of rank 1 such that:*

1. $T_i^2 \neq 0$,
2. $T_iT_j = 0$ *for* $j < i$.

Then $S = T_1 + \ldots + T_k$ *is of rank k and the null space of* S *is the intersection of the null spaces of the* T_i's.

Proof. Since T_i is of rank 1, $VT_i = Kv_i$ where $v_iT_i \neq (0)$ (since $T_i^2 \neq 0$) and $v_iT_j = 0$ for $j < i$ (since $T_iT_j = 0$ for $j < i$). From this it is trivial that $v_1, \ldots, v_k$ are linearly independent over K and that VS is the subspace spanned by the v_i's. If $vS = 0$ then $vST_k = 0$ giving $vT_k = 0$. Also $vST_{k-1} = 0$, and since $vT_k = 0$ this gives $vT_{k-1} = 0$. Continuing we get $vT_i = 0$ for $i = 1, 2, \ldots, k$.

Let φ be the isomorphism imbedding R into Q, the ring of K-linear transformations on V. It is clear that $a \in R$ is regular if and only if $\varphi(a)$ is invertible in Q.

Basic to our needs is the

LEMMA 4.5. *If* I *meets all left ideals of* R *then* I *contains a regular element.*

Proof. If $x \in R$ is such that $\varphi(x) \in Q$ has rank 1 then $axb = 0$ forces $ax = 0$ or $xb = 0$. We will find n elements $u_1, \ldots, u_n$ in I such that:

1. $\varphi(u_i)$ is of rank 1 for each i
2. $u_i^2 \neq 0$
3. $u_i u_j = 0$ for $j < i$.

We proceed by induction, supposing that we have $u_1, \ldots, u_k$, $k < n$ with the desired properties. There is a $u \neq 0$ in F_r so that $j(u)\varphi(u_i) = 0$ for $i = 1, 2, \ldots, k$ since there is a $j(u)$ annihilated by $\varphi(u_1) + \ldots + \varphi(u_k)$ (which is of rank less than n, so by the Sublemma B the result holds). Since $u \in F_r$, $\varphi(u)$ has rank 1. Also $Ru \neq (0)$ thus must intersect I; by the primeness of R, $u(Ru \cap I) \neq (0)$. But then there is $a = ru \neq 0$ in $Ru \cap I$ such that $uru \neq 0$. If $a^2 = 0$ then $ruru = 0$, whence, since $\varphi(u)$ is of rank 1, $ru = 0$ or $uru = 0$; at any rate uru would result contrary to choice. Let $u_{k+1} = a$; then $a^2 \neq 0$ so $u_{k+1}^2 \neq 0$, $\varphi(u_{k+1})$ is of rank 1, u_{k+1} is in I and since $j(u)$ is annihilated by all the $\varphi(u_i)$, $i = 1, 2, \ldots, k$ and $a = ru$, $j(u_{k+1})\varphi(u_j) = 0$ for all $j \leq k$. We have established the existence of $u_1, \ldots, u_n$ with the requisite properties.

But then $\varphi(u_1 + \ldots + u_n)$ is of rank n so is invertible in Q. Therefore $u_1 + \ldots + u_n$ is regular and is in I. This proves the lemma.

We pass to the first Goldie theorem.

THEOREM 4.4. <u>The ring</u> Q <u>of all</u> K-<u>linear transformations on</u> V <u>is a left quotient ring for</u> R. <u>In other words, a prime Goldie ring has a left quotient ring which is a simple ring with descending chain conditions on left ideals</u>.

<u>Proof.</u> We already know that a regular element in R picks up an inverse in Q. It remains to show that if $m \in Q$ then $m = \varphi(a)^{-1}\varphi(b)$ where $a, b \in R$ and a is regular.

Let $T = \{x \in R \mid \varphi(x)m \in \varphi(R)\}$; T is clearly a left ideal of R. We claim that T meets all left ideals of R. Let $I \neq (0)$ be a left ideal of R and let $u \neq 0 \in I \cap F_r$. The element $j(u)m \in V$ so we know that $j(u)m = \lambda_v^{-1} j(w)$ for suitable $v, w \in F_r$. Thus $\lambda_v j(u)m = j(w)$ giving us $j(vu)m = j(w)$. But then for $x \in F_r$, $j(x)\varphi(vu)m = \lambda_x j(vu)m = \lambda_x j(w) = j(x)\varphi(w)$, that is $\varphi(vu)m = \varphi(w)$. Therefore $vu \in T$ and since $\lambda_v \neq 0$, $vu \neq 0$. The element vu is in $T \cap I$. By Lemma 4.5, T has a regular element a, hence $\varphi(a)m \in \varphi(R)$ that is, $\varphi(a)m = \varphi(b)$ therefore $m = \varphi(a)^{-1}\varphi(b)$. This proves the theorem.

We prove a converse to Goldie's theorem. However since we shall need a wider result for later use we now prove this more general result.

DEFINITION. R is said to be <u>left-order</u> in S if S is a ring of left quotients for R.

THEOREM 4.5. <u>Let</u> R <u>be a left-order in</u> S <u>where</u> S <u>is a semi-simple ring with descending chain condition. Then</u> R <u>is a Goldie ring having no non-zero nilpotent ideals. Moreover, if</u> S <u>is simple then</u> R <u>is prime</u>.

Proof. We first show that R is a Goldie ring. Since S is semi-simple with descending chain conditions then, as is well-known, S must satisfy the ascending chain condition on left-ideals, hence in particular the ascending chain conditions on left annihilators. Since this last condition is inherited by subrings we see that R satisfies condition 1 of the Goldie conditions.

Let $\lambda_1, \ldots, \lambda_k$ form a direct sum of left ideals of R; we claim that $S\lambda_1, \ldots, S\lambda_k$ form a direct sum of left ideals in S. If $s_1\alpha_1 + \ldots + s_k\alpha_k = 0$ where $s_i \in S$, $\alpha_i \in \lambda_i$ then, as in Sublemma A, we can write each $s_i = s^{-1}b_i$, $s, b_i \in R$, s regular, thus $b_1\alpha_1 + \ldots + b_k\alpha_k = 0$. By the directness of the sum in R we get each $b_i\alpha_i = 0$ hence $s_i\alpha_i = 0$. Therefore R is a Goldie ring.

Let $N \neq (0)$ be a nilpotent ideal in R with $N^m = (0)$, $N^{m-1} \neq (0)$. The ideal SNS of S is not (0); since S is a semi-simple ring with minimum condition SNS = eS where e is a central idempotent contained in SNS. Now $e = \Sigma a_i u_i b_i$, $a_i, b_i \in S$ and $u_i \in N$; we can write each a_i as $a_i = a^{-1}c_i$, $a \in R$ is regular and $c_i \in R$. Thus $e = a^{-1}\Sigma c_i u_i b_i$ whence $e = a^{-1}\Sigma w_i b_i$, $w_i \in N$. Since e is central, ae = ea therefore $N^{m-1}ae = N^{m-1}\Sigma w_i b_i = (0)$; because a is regular this forces $N^{m-1}e = (0)$. Therefore $(N^{m-1}S)^2 = N^{m-1}SN^{m-1}S \subset N^{m-1}eS = (0)$. Since S is semi-simple this is impossible unless $N^{m-1} = (0)$. This contradiction proves that N = (0).

Finally, suppose that S is simple. Let $A \neq (0)$ be an ideal of R; thus SAS = S. Since S has a unit element 1, $1 = \Sigma b_i a_i c_i$, $b_i, c_i \in S$, $a_i \in A$. As before $b_i = b^{-1}d_i$, b regular, $d_i \in R$, thus $b = \Sigma d_i a_i c_i \in AS$. If for some ideal B of R, BA = (0) then BAS = (0) whence Bb = (0). The regularity of b forces B = (0), consequently R is a prime ring.

We have all that we need to prove a theorem due to Faith and Utumi. The very simple, elegant proof we give is due to Claudio Procesi.

THEOREM 4.6. Let R be a left-order in D_n, the ring of $n \times n$ matrices over the division ring D. Then we can find an Ore domain Λ whose ring of quotients is isomorphic to D and such that R contains a subring isomorphic to Λ_n.

Proof. By the previous theorem R is a prime Goldie ring so has a ring of left quotients isomorphic to K_m; by Wedderburn's Theorem $m = n$ and $K \approx D$. We use our previous notation F_r, E, K, $V = K \otimes_E F_r$. Since $K, F_r \subset K_n$ we can identify V with KF_r. Let $u_1, \ldots, u_n \in F_r$ form a basis of V over K; by Lemma 4.4, $Ru_1 + \ldots + Ru_n$ meets all left ideals of R hence, by Lemma 4.5, it contains a regular element c, $c = \Sigma r_i u_i$. The elements $u_i c^{-1}$ still form a basis of V; let $\Lambda = \{\lambda \in E \mid \lambda u_i c^{-1} \in F_r \text{ all } i\}$. Since $v \in V$ can be written as $\lambda_a^{-1} w$, $a, w \in F_r$ we know that $\Lambda \neq (0)$. Now Λ is a left ideal of E so by Lemma 4.3 its ring of quotients is also K. Now Q contains the ring $S = \{q \in Q \mid u_i c^{-1} q \in \sum_j \Lambda u_j c^{-1}\}$ for all i (that is the matrices in the basis $u_i c^{-1}$ having entries in Λ). S is isomorphic to Λ_n.

We claim that $S \subset R$ since if $u_i c^{-1} q \in \Sigma \Lambda u_j c^{-1} \subset F_r$ (by definition of Λ) then $q = cc^{-1} q = (\Sigma r_i u_i) c^{-1} q = \Sigma r_i (u_i c^{-1} q) \in RF_r \subset R$. Thus $R \supset S$ and $S \approx \Lambda_n$.

We should like to extend Goldie's theorem from prime rings to a much wider context. We so so soon, but first a definition.

DEFINITION. A ring R is said to be semi-prime if it has no non-zero nilpotent ideals.

Until further notice R will denote a semi-prime Goldie ring.

Let $\mathcal{M}$ be the set of all two-sided ideals of R which are the left annihilators of non-zero two-sided ideals of R. If R consisted merely of (0) then R would be a prime ring so we would know its structure from Theorem 4.4. So we may assume that (0) is not maximal in $\mathcal{M}$.

As a set of left annihilators $\mathcal{M}$ has maximal elements; we shall denote these by A or by A with a subscript. Our first task is to study these. But first we note: if A is maximal in $\mathcal{M}$ and r(A) is the right-annihilator of A then A is precisely the left annihilator of any non-zero ideal of R which is contained in r(A). In particular, if $xr(A) = 0$ then $x \in A$.

LEMMA 4.6. _If_ A _is maximal in_ $\mathcal{M}$ _then_ A _is a prime ideal of_ R.

Proof. Suppose that U,V are ideals of R such that $UV \subset A$; thus $UVr(A) = (0)$. If $V \not\subset A$ then $Vr(A) \neq (0)$ is an ideal of R which is contained in r(A). Since U annihilates Vr(A), U must lie in A. Therefore A is a prime ideal of R.

COROLLARY. _If_ $A_1 \neq A_2$ _are maximal in_ $\mathcal{M}$ _then_ $r(A_1) \subset A_2$.

LEMMA 4.7. _If_ $A_1, \ldots, A_n$ _are maximal in_ $\mathcal{M}$ _then_ $r(A_1) + \ldots + r(A_n)$ _is a direct sum. In particular,_ $\mathcal{M}$ _has only a finite number of maximal elements._

Proof. Consider $B = r(A_i) \cap (r(A_1) + \ldots + r(\hat{A}_i) + \ldots + r(A_n))$; then B annihilates $A_i + \prod_{j \neq i} A_j$ from the right. Since the A's are prime ideals, $\prod_{j \neq i} A_j \not\subset A_i$, so $A_i + \prod_{j \neq i} A_j$ annihilates B and is larger than A_i. Therefore by the choice of the A's, $B = (0)$. Thus the sum $r(A_1) + \ldots + r(A_n)$ is direct.

Since R is a Goldie ring it has no infinite direct sums of ideals, hence there are only a finite number of distinct $r(A_i)$ and so a finite number of maximal elements in $\mathfrak{M}$.

Let $A_1, \ldots, A_n$ be a complete list of the maximal elements in $\mathfrak{M}$. We want to show that $A_1 \cap A_2 \cap \ldots \cap A_n = (0)$. Before doing this we first make a remark.

<u>Remark</u>: In any semi-prime ring, if A, B are 2-sided ideals such that $AB = (0)$ then $BA = (0)$.

To see this merely note that $(BA)^2 = BABA = (0)$ since $AB = (0)$. Thus $BA = (0)$.

LEMMA 4.8. $A_1 \cap A_2 \cap \ldots \cap A_n = (0)$.

<u>Proof</u>. Let $W = A_1 \cap \ldots\ldots \cap A_n$. Now $W(r(A_1) + \ldots + r(A_n))$ $= (0)$ consequently by the above remark $(r(A_1) + \ldots + r(A_n))W = (0)$. If $W \neq (0)$ then $r(A_1) + \ldots + r(A_n)$ would annihilate a non-zero ideal of R from the left hence would be contained in an element of $\mathfrak{M}$ and so in a maximal element of $\mathfrak{M}$. In short $r(A_1) + \ldots + r(A_n) \subset A_j$ for some j; since $r(A_j) \subset r(A_1) + \ldots + r(A_n) \subset A_j$ we get $r(A_j)^2 \subset A_j r(A_j) = (0)$, forcing $r(A_j) = 0$, a contradiction. Therefore $W = (0)$.

We now have an imbedding of R into $R/A_1 \oplus \ldots \oplus R/A_n$ each of whose constituents, by Lemma 4.6, is a prime ring. We want -- in fact need -- that these are themselves Goldie rings. It is towards this goal we proceed, but first we establish a lemma which holds for any ring.

LEMMA 4.9. <u>Let</u> R <u>be any ring and suppose that</u> $A = \ell(S)$ <u>happens to be a two-sided ideal of</u> R. <u>Then if</u> $\overline{\lambda} \subset \overline{R} = R/A$ <u>is a left annihilator in</u> $\overline{R}$, <u>the left ideal</u> $\lambda = \{x \in R \mid \overline{x} \in \overline{\lambda}\}$ <u>is also a left annihilator in</u> R.

<u>Proof</u>. Suppose that $\overline{\lambda} = \ell(\overline{T})$; let T be the inverse image of $\overline{T}$ in R. Since $\overline{T} \neq (0)$ (otherwise $\overline{\lambda} = \overline{R}$, $\lambda = R$), $T \not\subset A$ so $TS \neq (0)$. We claim that $\lambda = \ell(TS)$. Clearly $\lambda TS = (0)$ since $\overline{\lambda}\,\overline{T} = (0)$, hence $\lambda T \subset A$. Also, if $xTS = (0)$ then $xT \subset A$ so that $\overline{x}\,\overline{T} = (0)$, putting $\overline{x} \in \overline{\lambda}$. This proves the lemma.

We return to our semi-prime Goldie ring.

LEMMA 4.10. <u>If</u> A <u>is maximal in</u> $\mathfrak{M}$ <u>then</u> R/A <u>is a Goldie ring.</u>

<u>Proof.</u> By Lemma 4.7 there is a correspondence between left annihilators in R/A (since $A = \ell(r(A))$) and certain left annihilators in R. Thus the ascending chain condition on left annihilators in R implies that in R/A.

What about direct sums of left ideals? Let $\overline{\lambda}_1, \ldots, \overline{\lambda}_k$ in R/A form a direct sum of left ideals, and let $\lambda_1, \ldots, \lambda_k$ be their inverse images in R. Hence $\lambda_i \cap (\lambda_1 + \ldots + \hat{\lambda}_i + \ldots + \lambda_k) \subset A$. Since $\lambda_i \not\subset A$, $\lambda_i r(A) \neq (0)$ therefore by the remark made earlier $r(A)\lambda_i \neq (0)$. Let $\mu_i = r(A)\lambda_i$; these are non-zero left ideals of R, and are contained in the corresponding $\lambda_i \cap r(A)$. However, $\tau_i = \mu_i \cap (\mu_1 + \ldots + \hat{\mu}_i + \ldots + \mu_k) \subset \lambda_i \cap (\lambda_1 + \ \ + \hat{\lambda}_i + \ldots + \lambda_k) \subset A$; τ_i also being in $r(A)$ we get it in $A \cap r(A) = (0)$. The μ_i thus give us a direct sum of non-zero left ideals in R of the same length as that given us by the $\overline{\lambda}_i$'s. Since there are no infinite direct sums of non-zero left ideals in R we get the same conclusion in R/A, that is, R/A is a Goldie ring.

We have all the requisite pieces to prove the second Goldie theorem.

THEOREM 4.7. <u>Let</u> R <u>be a semi-prime Goldie ring; then</u> R <u>has a ring of left quotients which is a semi-simple ring with descending chain conditions on left ideals.</u>

Proof. We already know that we have a finite set of prime ideals $A_1, \ldots, A_n$ such that each $R_i = R/A_i$ is a prime Goldie ring and such that R is isomorphically imbeddable in $S = R_1 \oplus \ldots \oplus R_n$. Moreover, by Theorem 4.4 each R_i has a ring of left quotients $Q(R_i)$ which is a total matrix ring over a division ring D_i.

We claim that $T = Q(R_1) \oplus \ldots . \oplus Q(R_n)$ is a ring of left quotients of R.

Now for any $j \neq i$, by the corollary to Lemma 4.6, $r(A_i) \subset A_j$ so its image in S lies in $r(A_i)$, is not (0) and is a two-sided ideal of R_i. Since R_i is prime, the image of $r(A_i)$ in R_i must meet every left ideal of R_i, thus by Lemma 4.5 it contains a regular element $\overline{c}_i$. Let $c_i \in r(A_i)$ be an inverse image of $\overline{c}_i$; the element $(c_1, c_2, \ldots, c_n)$ is the image in S of $c_1 + c_2 + \ldots + c_n$, and so, quite easily, is regular. But then $(c_1, \ldots, c_n)$ is invertible in T.

Let $T' = \{a^{-1}b \mid a, b \in R,\ a \text{ invertible in } T\}$. From the above $e_i = (0, 0, \ldots, 1, 0, \ldots, 0) = (c_1, \ldots, c_n)^{-1}(0, \ldots, c_i, 0, \ldots, 0)$ is in T' (we have identified $c_1 + \ldots + c_n$ with its image in S) consequently $T' \supset R_i$ for each i. But then $T' = T$. We have proved that T is a ring of left quotients for R. This proves the second Goldie theorem.

We now go to the third Goldie theorem wherein we shall investigate the nature of rings in which every left ideal is principal. The proof we give here is due to Procesi.

LEMMA 4.11. *Let R be a prime ring in which every left ideal is principal. Then R is isomorphic to a total matrix ring over an Ore domain.*

Proof. We use the notation used earlier in proving Theorem 4.4. If $1 \otimes u_1, \ldots, 1 \otimes u_n$ is a basis of V over K, $u_i \in F_r$, we have seen in Lemma 4.4 that $T = Ru_1 + \ldots + Ru_n$ meets all left ideals of R, thus by Lemma 4.5 it contains a regular element d. Since every left ideal is principal $T = Rc$ and so $d = rc$. The regularity of d implies that of c, hence c is regular. Moreover, $c = \Sigma b_i u_i$ being in $T = Ru_1 + \ldots + Ru_n$. Since each $u_i \in T$, $u_i = r_i c$ therefore $c = \Sigma b_i r_i c = (\Sigma b_i r_i)c$; since c is regular we deduce from this that $\Sigma b_i r_i = 1$. Since $u_i \in F_r$, $Fu_i = (0)$ so we get $Fr_i c = (0)$ therefore $Fr_i = (0)$, that is, each $r_i \in F_r$. We claim that $1 \otimes r_1, \ldots, 1 \otimes r_n$ form a basis of V over K, since the $1 \otimes (r_i c)$ do and c is regular. If $u \in F_r$ then $u = u1 = \Sigma ub_i r_i$, $ub_i \in F_r$ so that $u = \Sigma \lambda_{ub_i} j(r_i)$ and hence F_r is a free E-module of rank n with generators $j(r_1), \ldots, j(r_n)$.

Every element in R induces an E-endomorphism on F_r. Given an E-endomorphism of F_r, by Goldie's theorem (Theorem 4.4) it is inducible by right multiplication by $a^{-1}b$, $a, b \in R$ and a regular. Since $a^{-1}b$ is an E-endomorphism of F_r, each $r_i a^{-1} b$ is in F_r. But $a^{-1}b = 1a^{-1}b = (\Sigma b_i r_i)a^{-1}b = \Sigma b_i(r_i a^{-1}b)$ is an element of $RF_r \subset R$ since each $r_i a^{-1} b \in F_r$. Therefore $a^{-1}b \in R$ and the result is proved.

One could ask if the Ore domain in the statement of the lemma also has every left ideal principal. This is false as is shown by an example due to Swan.

LEMMA 4.12. A semi-prime ring whose left ideals are principal is a direct sum of a finite number of prime rings.

Proof. In the proof of Theorem 4.7 it is enough to show that $A_i + A_j = R$ for $i \neq j$ for then the result holds by the Chinese Remainder Theorem.

Now $A_i = Ra$, $A_j = Rb$ so $A_i + A_j = Rt$. The element t is regular for if $tx = 0$ then $A_i x = (0)$ and $A_j x = (0)$ so $x \in r(A_i) \cap r(A_j) = (0)$.

However, $t = xa + yb$, $a = ut$, $b = vt$ so $(xu + yv)t = t$ whence $xu + yv = 1$. If $I = \ell(Ra)$ then since $IA_i = (0)$, $A_i I = (0)$ by the semi-primeness of R, and so $I = r(Ra_i)$. But $IRa = (0)$ forces $IRut = (0)$ hence $IRu = (0)$, so $RuI = (0)$. Since $Ru \subset \ell(r(A_i)) = A_i$ we get $Ru \subset Ra$. Similarly $Rv \subset Rb$, so $1 \in Ru + Rv \subset Ra + Rb = A_i + A_j$. Thus $A_i + A_j = R$.

DEFINITION. R is a primary ring if R/N is a prime ring where N is the maximal nilpotent ideal of R.

LEMMA 4.13. A principal left ideal ring is the direct sum of primary rings.

Proof. Let N be the maximal nilpotent ideal of R. By Lemma 4.10, $R/N = \overline{R}_1 \oplus \ldots \oplus \overline{R}_k$ where the $\overline{R}_i$ are prime rings. We want to show that $R = R_1 \oplus \ldots \oplus R_k$ where R_i maps onto $\overline{R}_i$ modulo N.

Now $\overline{R} = R/N = \overline{A} \oplus \overline{B}$ where $\overline{A}$ is semi-prime and $\overline{B}$ is prime ($\overline{B} = R_k$, $\overline{A} = \overline{R}_1 \oplus \ldots \oplus \overline{R}_{k-1}$). Now $r(\overline{A}) = \ell(\overline{A}) = \overline{B}$ and $r(\overline{B}) = \ell(\overline{B}) = \overline{A}$ by the semi-primeness of $\overline{A}$ and $\overline{B}$. Let A, B be the inverse images of $\overline{A}$ and $\overline{B}$ respectively. Thus $A = Ra$, $B = Rb$ and $N \subset (Ra \cap Rb)$. If $xa \in N$ then $\overline{x}\overline{a} = 0$; but $\overline{R}\overline{a} = \overline{A}$ so $\overline{a}$ is regular in $\overline{A}$ which forces $\overline{x} \in \overline{B}$, hence $xa \in N$ implies that $x \in B$. Since $N \subset Ra$ we get $N \subset Rba$ (for $n \in N$, $n = xa$ forces $x \in B = Rb$) and $ba \in N$ since $\overline{b}\,\overline{a} = 0$. Thus $N = Rba$.

Also $N = Rab \subset RaRb \subset N$ so we find that $AB = BA = A \cap B = N$. We claim that for any integer $m > 0$, $R = A^m + B^m$; we go by induction. Now $R = R^2 = (A^m + B^m)(A + B) = A^{m+1} + B^{m+1} + U$ where $U \subset N$. Since $N \subset A$ and $N \subset B$ and $U \subset N = NR = N(A^m + B^m) \subset A^{m+1} + B^{m+1}$ we get $R = A^{m+1} + B^{m+1}$.

Since we have the ascending chain condition on left ideals in R there is an integer k such that $\ell(a^k) = \ell(a^{k+1}) = \dots$. But then $Ra^k \cap \ell(a^k) = (0)$ for if $xa^k \in \ell(a^k)$ then $0 = xa^k a^k = xa^{2k}$ hence $x \in \ell(a^{2k}) = \ell(a^k)$ giving us $xa^k = 0$.

Suppose that $N^t = (0)$; pick $m > \max(k, t)$. Then $B^m A^m = (BA)^m = (0)$ (since $BA = AB = N$); also $B^m \subset \ell(a^m)$ so $B^m \cap Ra^m = (0)$. Since $A = Ra$ is a 2-sided ideal, $aR \subset Ra$ hence $A^m = (Ra)^m \subset Ra^m \subset A^m$, thus $A^m = Ra^m$ and $B^m \cap A^m = (0)$.

Since $A^m \oplus B^m = R$ and since B^m maps on $\overline{B}$ and $\overline{A}^m$ maps on $\overline{A}$ modulo N we get the result.

We now prove the third Goldie theorem.

THEOREM 4.8. *Let R be a principal left ideal ring which satisfies the ascending chain condition on right ideals. Then R is the direct sum of a semi-prime ring in which all left ideals are principal and a ring with descending chain condition on left and right ideals.*

Proof. In light of the preceding lemmas, in order to prove the theorem it is enough to show that a primary but not prime ring of this type satisfies the descending chain conditions.

Let R be a primary ring and let $(0) \neq N = Rn$ be its maximal nilpotent ideal. We first want to show that R/N satisfies the descending chain conditions, or in other words (by Goldie's theorem) that R/N is its own ring of left quotients.

We claim that we can reduce to the case in which $N^2 = (0)$ for $R/N \approx (R/N^2)/(N/N^2)$ and R/N^2 is primary, not prime (if $N^2 \neq (0)$), in which $N^2 = (0)$. So we suppose that $N^2 = (0)$.

1. $r(N) = N$. For if $r(N) \neq N$ being an ideal in R which contains N, $r(N)/N \neq (0)$ is an ideal of R/N which is prime so meets all left ideals of R/N hence has a regular element. So there would be an element $c \in R$ with $Nc = (0)$ where $\overline{c}$ is regular in $\overline{R} = R/N$. Since $\overline{c}$ is not nilpotent neither is c and there is an integer k such that $\ell(c^k) \cap Rc^k = (0)$. Thus since $N \subset \ell(c^k)$, $N \cap Rc^k = (0)$. Now $N + Rc^k$ is a left ideal so $N + Rc^k = Rd$, $c^k = sd$, $n = rd$, $an + bc^k = d$. Therefore in $\overline{R}$, $\overline{d}$ is regular (since $\overline{c}^k$ is) and since $\overline{r}\overline{d} = 0$, $\overline{r} = 0$ so $r \in N$. Thus $ran + rbc^k = rd = n$; since $ran \in N^2 = (0)$ and $rb \in N$, $rbc^k = 0$ we get $n = 0$, a contradiction.

2. If $\overline{c}$ is regular in $\overline{R}$ then $Nc = N = Nc^k$. For if $c \to \overline{c}$ then $Rn + Rc = Rd$ where $\overline{d}$ is regular in $\overline{R}$ (since $\overline{c} = \overline{sd}$). Now $n = td$ and $d = pn + qc$; since $\overline{td} = 0$, $t \in N$ since $\overline{d}$ is regular. So $n = td = tpn + tqc = tqc$ since $tpn \in N^2 = (0)$. Thus $N \subset Nc \subset N$ proving $Nc = N$, and so also $Nc^k = N$.

3. If $\overline{c}$ is regular in $\overline{R}$ then c is left-regular in R, that is, $\ell(c) = (0)$. For we have $N = Nc^k$; pick k so that $Rc^k \cap \ell(c^k) = (0)$. Since $\overline{c}$ is regular, $\ell(c^k) \subset N = Nc^k$, hence $\ell(c^k) = (0)$. Thus c^k, and so, c is left-regular.

We finish the proof of the theorem. Let $\overline{c}$ be regular in $\overline{R}$ and pick $m \in N$ such that $Rm = N$ and mR is maximal over all such m. Now since $Nc = N$, $m = m_1c$ with $m_1 \in N$. Also $Nc = N = Rm = Rm_1c$ hence, since c is left-regular, $N = Rm_1$. Since $m = m_1c \in m_1R$, by the maximality of mR we must have $mR = m_1R$, that is, $m_1 = mx = m_1cx$, hence $m_1(1 - cx) = 0$.

Consequently $1 - cx \in r(N) = N$ whence in $\overline{R} = R/N$, $1 = \overline{cx}$ so $\overline{c}$ is invertible in $\overline{R}$. This proves $\overline{R} \approx D_n$ by Goldie's theorem so certainly satisfies the descending chain conditions.

We now must merely show that R also satisfies the descending chain conditions. Now $N^k = (0)$ and the R/N modules $N/N^2, N^2/N^3, \ldots, N^i/N^{i+1}$ are finitely generated. The nature of R/N from above then tells us N has a finite composition series, which together with the fact that R/N also has one tells us R satisfies the descending chain condition. A recent example due to Jategaonkar shows that one can not drop the assumption that R be right Noetherian.

We close this chapter with a fast simple and self-contained proof of Goldie's theorem for semi-prime rings recently worked out jointly by Procesi and Small.

For the first lemma we merely need that R be semi-prime with ascending chain condition on left annihilators and not the whole assumption that R be Goldie.

LEMMA 4.14. _If_ $A \supset B$ _are left ideals of_ R _and_ $r(A) \subset r(B)$ _but_ $r(A) \neq r(B)$ _then there exists an_ $a \in A$ _such that_ $Aa \neq (0)$ _and_ $Aa \cap B = (0)$.

Proof. Let U be a right annihilator minimal with respect to being contained in r(B) and properly containing r(A). By the choice $AU \neq (0)$ and since there are no nilpotent ideals, $AUAU \neq (0)$. Pick $ua \in UA$ such that $AuaU \neq (0)$. We claim that $Aua \cap B = (0)$. If not, $0 \neq xua \in Aua \cap B$ where $x \in A$. Since $xua \in B$, and $U \subset r(B)$, $xuaU = (0)$. Since $x \in A$, $r(x) \supset r(A)$. Consider $r(x) \cap U$; it is a right annihilator, contains r(A) and is contained in U; moreover, since $uaU \subset r(x) \cap U$ and is not in r(A) we get that $r(x) \cap U = U$, that is, $U \subset r(x)$, so $xU = (0)$, contradicting that $xua \neq 0$. This proves the lemma.

DEFINITION. A left ideal I is <u>essential</u> if it intersects every non-zero left ideal non-trivially.

COROLLARY 1. *If* Rx *and* Ry *are essential then* Rxy *is essential.*

Proof. Let $A \neq (0)$ be a left ideal of R and let $\overline{A} = \{r \in R \mid ry \in A\}$. By the essentiality of Ry, $\overline{A}y = Ry \cap A \neq (0)$. By definition, $\overline{A} \supset \ell(y)$ and $\overline{A}y \neq (0)$ so we can apply Lemma 4.14 to $\overline{A}$ and $\ell(y)$ to get a left ideal $T \neq (0)$ such that $T \subset \overline{A}$, $T \cap \ell(y) = (0)$. Let $\overline{T} = \{r \in R \mid rx \in T\}$; thus $\overline{T}x = Rx \cap T \neq (0)$, and $\overline{T}xy \neq (0)$ since $T \cap \ell(y) = (0)$. Thus since $\overline{T}xy \subset Ty \subset \overline{A}y \subset A$ we get $Rxy \cap A \neq (0)$, proving that Rxy is essential.

COROLLARY 2. *If* Ra *is essential then* a *is regular.*

Proof. That $r(a) = (0)$ follows from the special case of Lemma 4.14 in which $A = R$ and $B = Ra$.

We consider $\ell(a)$. We know there is an integer k such that $\ell(a^k) = \ell(a^{k+1}) = \ldots$. If $x \in Ra^k \cap \ell(a)$ then $x = ya^k$ and $0 = xa = ya^{k+1}$ placing $y \in \ell(a^{k+1}) = \ell(a^k)$, that is, $x = 0$. Since Ra^k is essential by Corollary 1 we get that $\ell(a) = (0)$.

From now on, R will be a semi-prime Goldie ring.

LEMMA 4.15. R *satisfies the descending chain condition on left annihilators.*

Proof. Let $L_1 \supset L_2 \supset \ldots \supset L_n \supset \ldots$ be a properly descending chain of left annihilators. By Lemma 4.14 there exist non-zero left ideals $C_n \subset L_n$ such that $C_n \cap L_{n+1} = (0)$. Since the C_n form a direct sum of left ideals we get that the chain of annihilators must terminate.

LEMMA 4.16. *If* $\ell(c) = (0)$ *then* Rc *is essential and so* c *is regular.*

Proof. If $A \neq (0)$ is a left ideal satisfying $Rc \cap A = (0)$, since $\ell(c) = (0)$, we get that the Ac^n form a direct sum of left ideals. Therefore $Rc \cap A \neq (0)$ and Rc is essential.

A two-sided ideal of S is said to be an *annihilator ideal* if S is the left annihilator of some left ideal T.

LEMMA 4.17. a) *A non-zero minimal annihilator ideal is a prime Goldie ring.*

b) *There is a finite direct sum of such ideals which is an essential left ideal of* R.

Proof. a) Let S be a minimal annihilator ideal. If $T \neq (0)$ is a left ideal of S, since R is semi-prime, $ST \neq (0)$; since $ST \subset T$ is a left ideal of R this immediately implies that S has no infinite direct sums of left ideals. It is immediate that the chain condition on left annihilators is inherited by subrings, hence S is a Goldie ring.

Suppose that $AB = (0)$ where A, B are ideals of S. Since $SB \subset B$, $ASB = (0)$, thus $A \subset \ell(SB) \cap S$; if $A \neq (0)$ then by the minimality of S we get $S \subset \ell(SB)$, and since $SB \subset S$ this gives $SB = (0)$ and so $B = (0)$. In short, S is a prime Goldie ring.

b) Let $A = S_1 \oplus S_2 \oplus \ldots \oplus S_n$ be a maximal direct sum of minimal annihilator ideals. We claim that A is essential; suppose that $A \cap K = (0)$ where K is a non-zero left ideal of R. Since $AK \subset A \cap K = (0)$, $AK = (0)$, so $K \subset r(A)$. But since R is semi-prime, $A \cap r(A) = (0)$; since $r(A)A = (0)$ (since R is semi-prime) we can find a minimal, non-zero annihilator ideal in $r(A)$, and so lengthen the direct sum giving A. Thus A is essential.

LEMMA 4.18. *If* I *is an essential left ideal of* R *then it contains a regular element.*

Proof. We first establish the result for prime rings. Choose $a \in I$ so that $\ell(a)$ is minimal. If a is regular, we are done; if not, by Corollary 2 to Lemma 4.14, $Ra \cap J = (0)$ for some left ideal $J \neq (0)$. Since I is essential, $I \cap J \neq (0)$ hence we may suppose that $J \subset I$. If $x \in J$, $r \in \ell(a+x)$, $r(a+x) = 0$ hence $ra = -rx \in Ra \cap J = (0)$ thus $r \in \ell(a) \cap \ell(x)$. By the minimality of $\ell(a)$ we get $\ell(a) \subset \ell(x)$ for all $x \in J$, thus $\ell(a)J = (0)$. Since $J \neq (0)$ and R is prime, $\ell(a) = (0)$ results. Lemma 4.16 then tells us that a is regular.

Now let R be semi-prime and let $A = S_1 \oplus \ldots \oplus S_n$ be as in Lemma 4.17. It is clear that $I \cap S_i$ is essential in S_i. Since S_i is prime, $I \cap S_i$ contains an element r_i regular in S_i. Let $r = r_1 + \ldots + r_n$; we claim that r is regular in R; if not $\ell(r) \cap A \neq (0)$ so we can find $t_1, \ldots, t_n \in S_1, \ldots, S_n$ respectively and not all 0 such that $tr = \Sigma t_i r_i = 0$. Then $t_i r_i = 0$ all i, violating the regularity of the r_i in S_i. The element r is the required regular element in I.

THEOREM 4.9. *Let* R *be a semi-prime Goldie ring. Then* R *has a ring of left quotients* Q.

Proof. Suppose that $a \in R$ is regular and $b \in R$. By Lemma 4.16, Ra is essential; moreover, as is easy to see, $M = \{r \in R \mid rb \in Ra\}$ is also essential, thus by Lemma 4.18, M contains a regular element c. Thus $cb = da \neq 0$; the condition for a left quotient ring to exist, as given in Theorem 4.1 holds in R so R has a ring of left quotients.

We remark two easy facts without proof.

1. If I is a left ideal of Q then $I = Q(I \cap R)$.

2. If $A_1 \oplus \ldots \oplus A_n$ is a direct sum of left ideals in R then $QA_1 \oplus \ldots \oplus QA_n$ is a direct sum in Q.

We now reprove Goldie's theorem (Theorem 4.7).

THEOREM 4.10. Q _is a semi-simple ring satisfying the descending chain condition on left ideals._

Proof. If $I \neq (0)$ is a left ideal of Q then there is a left ideal K in R such that $(I \cap R) \oplus K$ is essential in R (take K to be the complement in a largest direct sum containing $I \cap R$ as constituent). Since Q is a left quotient ring for R and $(I \cap R) \oplus K$ is essential in R we get that $Q = I \oplus QK$. If 1 is the unit element in Q, $1 = e + f$ where $e \in I$, $f \in QK$. Thus $e = e^2 + ef$, giving $e^2 = e \neq 0$. Also $I = Ie = Qe$. Thus every non-zero left ideal in Q is generated by an idempotent. In particular, Q can have no non-zero nilpotent ideals, so is semi-prime. Since every left ideal of Q is principal, Q satisfies the ascending chain condition on all left ideals so has no infinite direct sums of left ideals.

Let $L \neq (0)$ be a left ideal of Q; as we have seen $L = Qe$, so $r(L) = r(Qe) = (1 - e)Q$, whence $\ell(r(L)) = \ell((1 - e)Q) = eQ = L$. Therefore every left ideal of Q is a left annihilator. However, by Lemma 4.15, Q satisfies the descending chain condition on left annihilators, _thus on all left ideals._ Since, as we have seen, Q has no nilpotent ideals and has descending chain condition it must be semi-simple.

Bibliography

1. C. Faith and Y. Utumi, "On Noetherian prime rings", Trans. Amer. Math. Soc. 114 (1965), pp. 53-60.

2. A. W. Goldie, "The structure of prime rings under ascending chain conditions", Proc. London Math. Soc. 8 (1958), pp. 589-608.

3. A. W. Goldie, "Semi-prime rings with maximum conditions", Proc. London Math. Soc. 10 (1960), pp. 201-220.

4. A. W. Goldie, "Rings with maximum conditions", Arch. Math. 13 (1962), pp. 213-221.

5. I. N. Herstein, "Sul teorema di Goldie", Rend. Accad. Lincei xxxv (1963), pp. 23 - 26.

6. C. Procesi, "Sopra un teorema di Goldie riguardante la struttura degli anelli primi con condizioni di massimo", Rend. Accad. Lincei, xxxiv (1963), pp. 372-377.

7. C. Procesi, "Sugli anelli principali ed un teorema di Goldie", Rend. Accad. Lincei, xxxvi (1964), pp. 804-807.

8. C. Procesi, "Su un teorema di Faith ed Utumi", Rend. Mat. Roma, 24 (1965), pp. 346-347.

9. C. Procesi and L. Small, "On a theorem of Goldie", Journal of Algebra, 2 (1965), pp. 80-84.

CHAPTER 5

MORE ON RINGS WITH ASCENDING CHAIN CONDITIONS

We begin this chapter with an exposition of some results obtained jointly with Lance Small. We shall study nil subrings of rings satisfying the ascending chain condition on left annihilators. Our objective is to show that, quite generally, these are nilpotent.

Until further notice R shall be a ring satisfying the ascending chain condition on left annihilators. It is appropriate to remind the reader that this is equivalent to the descending chain condition on right annihilators. Also, since the left annihilator in a subring is merely the intersection of this subring with the left annihilator in the entire ring, our condition is inherited by all subrings of R.

We begin with

LEMMA 5.1. *Let* R *be a ring satisfying the ascending chain condition on left annihilators. If* R *is nil then every non-zero homomorphic image of* R *contains a non-zero nilpotent ideal.*

Proof. Let $\overline{R}$ be a non-zero homomorphic image of R; thus $\overline{R}$ is isomorphic to R/M where $M \neq R$ is a two-sided ideal of R.

Let $\mathcal{M} = \{\ell(x) \mid x \in R, x \notin M\}$; by hypothesis $\mathcal{M}$ has a maximal element, that is, there is an element a in R and not in M such that $\ell(a)$ is maximal in $\mathcal{M}$.

For any x in R, $\ell(ax) \supset \ell(a)$; if, in addition, ax is not in M then by the maximality of $\ell(a)$ we are forced to conclude that $\ell(ax) = \ell(a)$. Suppose that $ar \notin M$; since ar is nilpotent there is an integer k such that $(ar)^k \in M$ but $(ar)^{k-1} \notin M$. However, ar is nilpotent ; thus there is a least positive integer n such that $(ar)^n = 0$. Since $(ar)^{k-1}$ is of the form ax and is not in M, $\ell((ar)^{k-1}) = \ell(a)$. Because $(ar)^{n-k+1}$ is in $\ell((ar)^{k-1})$ we are led to $(ar)^{n-k+1}a = 0$, and so $(ar)^{n-k+2} = (ar)^{n-k+1}ar = 0$. By our choice of n we must conclude that $n-k+2 \geq n$ and so $k \leq 2$. Consequently, $(ax)^2$ is in M for every x in R.

For any x in R, since $x+1$ is formally invertible, we have that $ax+a$ is not in M ; from this results $\ell(ax+a) = \ell(a)$ and the argument used above yields $(ax+a)^2$ is in M.

In $\overline{R} = R/M$ these translate into: $\overline{a} \neq 0$, $(\overline{a}\overline{x})^2 = 0$, $(\overline{a}\overline{x} + \overline{a})^2 = 0$ for all $\overline{x}$, which immediately yield that $\overline{a}^2 = 0$ and also $\overline{a}\overline{x}\overline{a} = 0$ for all $\overline{x}$ in $\overline{R}$. Thus $\overline{a}\overline{R}$ is a nilpotent right ideal of $\overline{R}$. If $\overline{a}\overline{R} = (0)$ then the left annihilator of $\overline{R}$, would be a non-zero nilpotent ideal. If $\overline{a}\overline{R} \neq 0$ is generates a nilpotent two-sided ideal. In this way the lemma has been established.

As an immediate corollary we have the basic result:

COROLLARY. _If_ R _is a nil ring satisfying the ascending chain condition on left annihilators then_ R _is locally nilpotent._

Proof. As is well known (see for instance (N)) R contains a maximal locally nilpotent ideal $L(R)$, the Levitzki radical of R, which contains all locally nilpotent ideals of R. Moreover, $L(R/L(R)) = (0)$, hence $R/L(R)$ certainly contains no nilpotent ideals.

If $L(R) \neq R$, by the lemma, $R/L(R)$ would have a non-zero nilpotent ideal, which it can not. Thus $R/L(R) = (0)$, hence $R = L(R)$. Therefore, R is locally nilpotent.

We continue the investigation of our class of nil rings with

LEMMA 5.2. Let R be a nil ring satisfying the ascending chain condition on left annihilators. Then there exists an element $x_o \neq 0$ in R such that $Rx_o = (0)$.

Proof. Let $\mathcal{M}$ be the set of finitely generated subrings in R and let $\mathcal{N} = \{r(S) \mid S \in \mathcal{M}\}$. By the remark previously made, there is an S_o in $\mathcal{M}$ such thar $r(S_o)$ is minimal in $\mathcal{N}$. Let $x \in R$; then S_1, the subring generated by S_o and x, is finitely generated. Since $S_o \subset S_1$, $r(S_o) \supset r(S_1)$, whence, by the minimality of $r(S_o)$ we conclude that $r(S_o) = r(S_1)$. However, $r(S_1) = r(S_o) \cap r(x)$; together with the above equality we conclude that $r(S_o) = r(S_o) \cap r(x)$, that is $r(S_o) \subset r(x)$ for every x in R. But then $Rr(S_o) = (0)$.

To finish the proof we note that since S_o is finitely generated by the corollary to Lemma 5.1, it must be nilpotent. Hence for some integer k, $S_o^k = (0)$, $S_o^{k-1} \neq (0)$. Since $(0) \neq S_o^{k-1} \subset r(S_o)$, thus $r(S_o) \neq (0)$ and we have proved the lemma.

LEMMA 5.3. If R satisfies the ascending chain condition on left (respectively, right) annihilators and if A is a two-sided ideal of R which happens to be a left annihilator in R then R/A satisfies the ascending chain condition on left (respectively, right) annihilators.

Proof. As we have previously observed the ascending chain condition on right annihilators translates into the descending chain condition on left annihilators. In view of the one-to-one correspondence set up by Lemma 4.7 between the left annihilators in R/A and certain left annihilators in R the lemma now becomes clear.

We are now in a position to prove the first of the theorems of this chapter.

THEOREM 5.1. _If R is a ring which satisfies the ascending chain conditions on left and right annihilators then any nil subring of R is nilpotent._

Proof. Since the ascending chain conditions on left and right annihilators are inherited from the ring by its subrings we may, without loss of generality, assume that R is nil. We wish to show that R is nilpotent. Let $T_k = \{x \in R \mid xR^k = (0)\}$. Since $T_1 \subset T_2 \subset \ldots \subset T_k \subset \ldots$ and since they are left annihilators, by our hypothesis there is an n such that $T_n = T_{n+1} = \ldots$. If $T_n = R$ then $R^{n+1} = (0)$ and we are done. Suppose $T_n \neq R$; then $\overline{R} = R/T_n \neq (0)$. By Lemma 5.3, $\overline{R}$ satisfies the ascending chain condition on right annihilators so by Lemma 5.2 (in its reflected form) there exists an $\overline{x} \neq 0$ in $\overline{R}$ such that $\overline{x}\overline{R} = (0)$. Therefore $xR \subset T_n$ (where x is an inverse image of $\overline{x}$ in R) and so $xRR^n = (0)$. Thus $xR^{n+1} = (0)$; from our choice of n this forces $xR^n = (0)$ and so $x \in T_n$. But then $\overline{x} = 0$ contrary to hypothesis. This contradiction proves that $\overline{R} = (0)$ and so $R = T_n$ and $R^{n+1} = (0)$.

If one examines the proof given for Theorem 5.1, one sees that a more general result has actually been proved, that only a part of the ascending chain condition on the right has been used. The exact theorem proved was:

THEOREM 5.1'. _Let R be a nil ring satisfying the ascending chain condition on left annihilators and such that the ascending chain of right annihilators $T_n = \{x \mid R^n x = (0)\}$ becomes stationary. Then R is nilpotent._

An example by Sasiada shows that one can not get by with just the ascending chain condition on left annihilators. Let R be the ring generated over the integers by the elements $x_1, x_2, \ldots, x_n, \ldots$ subject to the conditions $x_j x_i = 0$ for $j \geq i$. Then as can easily be checked, R is nil, satisfies the ascending chain condition on left annihilators yet is not nilpotent.

We now consider nil rings satisfying a polynomial identity. Using a very simple argument due to Posner (see the proof of Lemma 5.7) we have that any prime ring satisfying a polynomial identity over its centroid satisfies the ascending chain condition on left and right annihilators. Thus, by our Theorem 5.1, any nil subring of such a ring is nilpotent. We summarize this in

LEMMA 5.4. *Let* R *be a prime ring satisfying a polynomial identity over its centroid. Then any nil subring of* R *is nilpotent.*

The above result can also be derived as a consequence of the characterization given by Posner of prime rings satisfying polynomial identities. We shall treat this theorem of Posner later.

Let R be any nil ring satisfying a polynomial identity with properly conditioned coefficients (e.g., R may be an algebra over a field satisfying a polynomial identity with coefficients in the field, R may be any ring, the identity having as coefficients operators, some of which are invertible, R/P may be supposed to satisfy some polynomial identity, depending on P, for each prime ideal P of R, etc., etc.). A theorem of Kaplansky asserts that such a ring is locally nilpotent. We give a new, and very easy proof of this in the

COROLLARY 1. *A nil ring satisfying a polynomial identity is locally nilpotent.*

Proof. Let S be a finitely generated subring of R which we suppose is not nilpotent. Since S^k is also finitely generated, by Zorn's lemma there exists an ideal P of R maximal with respect to not containing any S^k. However, P is a prime ideal of R for if $AB \subset P$ with A, B ideals properly larger than P then $A \supset S^{k_1}$, $B \supset S^{k_2}$ whence $P \supset AB \supset S^{k_1+k_2}$, a contradiction. However, $\overline{R} = R/P$ is a prime ring satisfying a polynomial identity; since, in addition, it is nil, by the lemma it is nilpotent. This is inconsistent with $\overline{R}$ being prime. Thus no such P exists and we conclude that $S^k = (0)$ for some k.

THEOREM 5.2. *Let* R *be a nil Goldie ring; then* R *is nilpotent.*

Proof. Let $A = \{x \in R \mid Rx = (0)\}$ and let T be the torsion part of A. Of course, T is a two-sided ideal of R. We can write T as a direct sum of its primary components, T_p, where if $t \in T_p$ then $p^{m(t)}t = 0$ where p is a prime.

Let $V_o = (0)$ and $V_1 = \{x \in T_p \mid px = 0\}$. V_1 is an ideal of R and is a vector space over the field P of p elements. Since there are no infinite direct sums of left ideals in R and since every subgroup of V_1, which is an algebra over P, is a left ideal of R we must conclude that V_1 is finite-dimensional over P. Let $V_2 = \{x \in T_p \mid p^2x = 0\}$, and let $W_2 = V_2/V_1$. If $u_1, \ldots, u_k \ldots$ are in V_2 then $pu_1, \ldots, pu_k \ldots$ are in V_1 which is finite-dimensional over P, hence $\Sigma a_i pu_i = 0$ for some non-zero a_i in P; hence $p\Sigma a_i u_i = 0$ and so $\Sigma a_i u_i \in V_1$. But then W_2 is finite-dimensional over P. Similarly, if $V_i = \{x \in T_p \mid p^i x = 0\}$ then $W_i = V_i/V_{i-1}$ is a finite-dimensional vector space over P.

Each V_i is a right ideal of R, hence V_i/V_{i-1} is an R-module which is a finite-dimensional vector space over P.

Since R induces a nil finite ring of linear transformations on these, this ring of linear transformations must be nilpotent. Therefore $V_i R^{n_i} \subset V_{i-1}$ for some integer n_i. From this we get that $V_i R^{m_i} = (0)$ where $m_i = n_1 + n_2 + \ldots + n_i$.

However, by the ascending chain condition on left annihilators there exists an integer k_o such that if $xR^m = (0)$ then $xR^{k_o} = (0)$. Given any element a in T_p it is in some V_i hence $aR^{m_i} = (0)$ and so $aR^{k_o} = (0)$. In other words, $T_p R^{k_o} = (0)$.

Since T is a direct sum of the T_p's we get that $TR^{k_o} = (0)$.

Consider $\overline{R} = R/T$ and let $\overline{A} = A/T$; $\overline{A}$ is torsion free. Let $V = \overline{A} \otimes_Z Q$ where Z is the ring of integers and Q is the field of rational numbers. If $\overline{u}_1, \ldots, \overline{u}_n$ in $\overline{A}$ are linearly independent (as elements in $\overline{V}$) over Q they are <u>a priori</u> linearly independent over Z, hence their inverse images $u_1, \ldots, u_n$ in A are linearly independent over Z. Since these then generate a direct sum of left ideals of R (since any subgroup of A is a left ideal of R) we get, from our hypothesis, that $\overline{V}$ is finite-dimensional over Q. Thus $\overline{R}$ induces a ring of linear transformations on $\overline{V}$ by $(\overline{a} \times q)t_{\overline{r}} = \overline{a}\overline{r} \times q$. The homomorphism $\overline{r} \to t_{\overline{r}}$ has kernel exactly $r(\overline{A})$, thus $\overline{R}/r(\overline{A})$ is a nil ring of $n \times n$ matrices over the rationals, where $n = \dim_Q \overline{V}$. By invoking our Theorem 5.1, $\overline{R}/r(\overline{A})$ is nilpotent, hence $(\overline{R}/r(\overline{A}))^m = (0)$, so $\overline{R}^m \subset r(\overline{A})$ and so $\overline{A}\,\overline{R}^m = (0)$, whence $AR^m \subset T$. Since $TR^{k_o} = (0)$ we get $AR^m R^{k_o} = (0)$ and so $AR^{m+k_o} = (0)$. By our choice of k_o it results that $AR^{k_o} = (0)$.

Let $U = \{x \in R \mid xR^{k_o} = (0)\}$. In $R' = R/U$ we know, by Lemma 5.3 that the ascending chain condition on left annihilators holds. If $R' \neq (0)$, since it is nil, we would have by Lemma 5.2

that there is an ideal $B' \neq (0)$ so that $R'B' = (0)$. Thus if B denotes the inverse image in R of B', $RB \subset U$, whence $RBR^{k_o} = (0)$. However, this places BR^{k_o} in A, and the discussion above allows us to conclude that $(BR^{k_o})R^{k_o} = (0)$. Since $BR^{2k_o} = (0)$, by the choice of k_o we are led to $BR^{k_o} = (0)$ and so $B \subset U$ and $B' = (0)$. This last conclusion is in variance with $B' \neq (0)$, forcing us to conclude that $R' = (0)$, that is, $R = U$. From the definition of U we obtain $R^{k_o+1} = (0)$ and so R is nilpotent. This proves the theorem.

Lanski has generalized Theorem 5.2 to the following: any nil subring of a Goldie ring is nilpotent. His method of proof is to show that the example of Sasiada cited right after Theorem 5.1' can not be imbedded in a Goldie ring. Using this, given a nil but not nilpotent subring of a Goldie ring, he constructs in this subring a sequence of elements that behave like those in Sasiada's example. This enables Lanski to arrive at his result.

We prove another theorem in the vein of nil implies nilpotent. In this theorem we suppose that R is an algebra over an integral domain Ω, that is, $\alpha x = 0$, $\alpha \in \Omega$, $x \in R$ implies that $\alpha = 0$ or $x = 0$ and $\alpha(xy) = (\alpha x)y = x(\alpha y)$.

THEOREM 5.3. _Let_ R _be a nil ring satisfying the ascending chain condition on left annihilators. If_ R _satisfies a polynomial identity (with coefficients in_ Ω_) then_ R _must be nilpotent._

Proof. By the ascending chain condition on left annihilators there is an integer k such that $xR^t = (0)$ then $xR^k = (0)$. Let $U = \{x \in R \mid xR^k = (0)\}$ and consider $\overline{R} = R/U$. If $\overline{x}\overline{R} = (0)$ then $xR \subset U$ so $xR^{k+1} = (0)$ resulting in $xR^k = (0)$, $x \in U$ and so $\overline{x} = 0$. If $U = R$ then R is nilpotent; if $U \neq R$ since ΩU clearly is in U, $\overline{R}$ is a nil ring in which $\overline{x}\overline{R} \neq (0)$ for $\overline{x} \neq 0$ and satisfies the same identity as R. All in all we have shown that if R

is not nilpotent we may assume that $xR = (0)$ only if $x = 0$. We proceed to show that this situation is impossible.

As is well known (and easy) we can suppose that the identity satisfied by R is of the form

$$p(x_1, \ldots, x_n) = \sum_{\sigma \in S_n} \alpha_\sigma x_{\sigma(1)} \cdots x_{\sigma(n)}$$

where $\alpha_\sigma \in \Omega$, S_n is the symmetric group of degree n.

Let $A = \{x \in R \mid Rx = (0)\}$; by Lemma 5.2, $A \neq (0)$. If $x_1 = a \in A$, $p(a, x_2, \ldots, x_n) = 0$ gives us, since $Ra = (0)$, that

$$a \sum_{\sigma \in S_{n-1}} \alpha_\sigma x_{\sigma(2)} \cdots x_{\sigma(n)} = 0$$

for $x_i \in R$, that is

$$A(\sum_{\sigma \in S_{n-1}} \alpha_\sigma x_{\sigma(2)} \cdots x_{\sigma(n)}) = (0).$$

Since $\Omega r(A) \subset r(A)$, $\overline{R} = R/r(A)$ satisfies the identity

$$q(x_2, \ldots, x_n) = \sum_{\sigma \in S_{n-1}} \alpha_\sigma x_{\sigma(2)} \cdots x_{\sigma(n)}$$

of degree $n-1$. By induction, since $\overline{R}$ is nil, satisfies the ascending chain condition on left annihilators (Lemma 5.3) and a polynomial identity of degree $n-1$ or less we conclude that $\overline{R}$ is nilpotent, say $\overline{R}^m = (0)$. Therefore $R^m \subset r(A)$ leading to $AR^m = (0)$. Since, as we have seen above, $\ell(R) = (0)$ we get the contradiction $A = (0)$. The theorem is now proved.

Let us recall something done in (N) about identities. The standard identity of degree k is defined by

$$[x_1, \ldots, x_k] = \Sigma\, (-1)^\sigma x_{\sigma(1)} \cdots x_{\sigma(k)}$$

where $(-1)^\sigma$ is the signature of the permutation σ in S_k and the sum runs over all $\sigma \in S_k$. If an algebra A is of dimension

k over a field F then it satisfies the standard identity $[x_1, \ldots, x_{k+1}]$. Thus the $n \times n$ matrices over a field satisfies $[x_1, \ldots, x_{n^2+1}]$. It is a fact, first proved by Amitsur and Levitski that they indeed satisfy $[x_1, \ldots, x_{2n}]$ and that this is the minimal identity. However we shall not need so sharp a result here. For a field F let $r(n)$ be the degree of the minimal identity satisfied by F_n. Following a very easy argument by Kaplansky we show

LEMMA 5.5. $r(n) \geq r(n-1) + 2$.

<u>Proof</u>. We imbed F_{n-1} in F_n as $\begin{pmatrix} F_{n-1} & 0 \\ 0 & 0 \end{pmatrix}$. Let $t = r(n-1) - 1$; hence we can find $a_1, \ldots, a_t \in F_{n-1}$ such that $[a_1, \ldots, a_t] \neq 0$. As imbedded in F_n we see immediately that $[a_1, \ldots, a_t, e_{kn}, e_{nn}] = [a_1, \ldots, a_t]e_{kn}e_{nn} = [a_1, \ldots, a_t]e_{kn} \neq 0$ for some k (since $[a_1, \ldots, a_t] = \begin{pmatrix} A & 0 \\ 0 & 0 \end{pmatrix}$ where $A \neq 0$). Thus indeed $r(n) > t+2 = r(n-1)+1$.

From this we clearly have that $r(n) \geq 2n$. We proceed to a result about finite-dimensional simple algebras.

THEOREM 5.4. <u>Let</u> R <u>be a simple algebra finite-dimensional over its center</u> F; <u>let</u> $A \subset R$ <u>be a subalgebra over</u> F <u>such that any polynomial identity satisfied by</u> A <u>over the prime field</u> P <u>of</u> F <u>is satisfied by</u> R. <u>Then</u> $A = R$.

<u>Proof</u>. Let $\overline{F}$ be the algebraic closure of F; the hypothesis carries over to $R \otimes_F \overline{F}$ and $A \otimes_F \overline{F}$. Hence to prove the theorem we may assume, without loss of generality, that F is algebraically closed. Thus by Wedderburn's theorem $R = F_n$.

If A is semi-simple then by Wedderburn's theorem $A \approx F_{n_1} \oplus \ldots \oplus F_{n_k}$ hence A satisfies all the identities satisfied over P by F_{n_t} where $n_t = \max(n_1, \ldots, n_k)$ thus by assumption F_n satisfies all the polynomial identities over P satisfied by F_{n_t}. By Lemma 5.10 we deduce that $n_t = n$ and hence $A = F_n$ the desired conclusion.

Assume then that A is not semi-simple and that its radical N is not (0). Now $A/N \approx F_{n_1} \oplus \ldots \oplus F_{n_k}$ where each $n_i < n$ since $\dim A/N < \dim A \leq \dim F_n$. If $m = \max(n_1, \ldots, n_k)$ then A/N satisfies the identities of F_m hence $[x_1, \ldots, x_r]$ where $r = r(m)$, that is to say, $[a_1, \ldots, a_r] \in N$ for all $a_i \in A$. Now N is nilpotent, say $N^t = (0)$, and is an ideal of A, hence satisfies the identity over P, $([x_1, \ldots, x_r]y)^t$. By hypothesis, F_n must satisfy this identity and since F_n is simple and has no nilpotent ideals we get F_n must satisfy $[x_1, \ldots, x_r]y$ and so, $[x_1, \ldots, x_r]$. Since $r = r(m)$ and $m = \max n_i < n$ we get a contradiction. This proves the theorem.

We now turn to a slightly different direction; we intend to introduce the idea of an ultra-product. While this has been used a great deal in logic it has not, until quite recently in the work of Amitsur, been used in ring theory. Much of the material to follow is treated along lines laid out by Amitsur.

Let $A = \{\alpha\}$ be a set; a set $\mathcal{F} = \{S\}$ of subsets of A is a filter if:

1. ϕ, the empty set, is not in
2. $S_1, S_2 \in \mathcal{F}$ implies $S_1 \cap S_2 \in \mathcal{F}$
3. $S \in \mathcal{F}$ and $T \supset S$ then $T \in \mathcal{F}$

A filter is said to be an ultra-filter if it is maximal in the natural ordering of filters: $\mathcal{F}_1 \leq \mathcal{F}_2$ if $S \in \mathcal{F}_1$ implies $S \in \mathcal{F}_2$. By Zorn's lemma it is clear that any filter can be enlarged to an ultra-filter.

Ultra-filters have the very simple characterization given by

LEMMA 5.6. A filter $\mathcal{F}$ on A is an ultra-filter if and only if for any subset T of A either $T \in \mathcal{F}$ or the complement of T, $A - T \in \mathcal{F}$.

Proof. If $\mathcal{F}$ is a filter and T is a subset of A, $T \notin \mathcal{F}$ then only one of the following can hold: $T \cap S = \phi$ for some $S \in \mathcal{F}$ or $(A - T) \cap S' = \phi$ for some $S' \in \mathcal{F}$. Otherwise

$\phi \neq S \cap S' = A \cap S \cap S' = ((A - T) \cup T) \cap S \cap S' \subset ((A - T) \cap S') \cup (T \cap S) = \phi$.

If $\mathcal{F}$ is an ultra-filter and there exists $S' \in \mathcal{F}$ with $(A - T) \cap S' = \phi$ then $T \cap S \neq \phi$ for all $S \in \mathcal{F}$. Thus $\{T, \mathcal{F}\}$ forms a base of a filter which, by the maximality of $\mathcal{F}$ forces $T \in \mathcal{F}$. By the same argument, if $(A - T) \cap S \neq \phi$ for all $S \in \mathcal{F}$, $A - T \in \mathcal{F}$. This proves the result in one direction.

If, on the other hand, $\mathcal{F}$ is a filter with the property that for any subset T of A either $T \in \mathcal{F}$ or $A - T \in \mathcal{F}$ then we claim that $\mathcal{F}$ is an ultra-filter. Suppose $\mathcal{F}'$ is a properly larger filter and $T \in \mathcal{F}'$, $T \notin \mathcal{F}$. Thus $A - T \in \mathcal{F}$ so $A - T \in \mathcal{F}'$, whence $\phi = T \cap (A - T) \in \mathcal{F}'$, a contradiction.

Let R_α be a set of rings indexed by $\alpha \in A$, and let $\prod_{\alpha \in A} R_\alpha$ be the complete product of the R_α's. As usual we can consider this ring as the set of functions with $f(\alpha) \in R_\alpha$ under pointwise addition and multiplication.

Let $\mathcal{F}$ be a filter on A; we define $f, g \in \prod R_\alpha$ to be equivalent if $\{\alpha \in A \mid f(\alpha) = g(\alpha)\}$ is in $\mathcal{F}$. This is indeed an equivalence relation $\sim$. Let $I = \{f \mid f \sim 0\}$; I is an ideal of $\prod R_\alpha$. We define $\prod R_\alpha/\mathcal{F}$ to be $\prod R_\alpha/I$. We call $\prod R_\alpha/\mathcal{F}$ an ultra-product if $\mathcal{F}$ is an ultra-filter.

A basic property of ultra-products is given in

THEOREM 5.5. *Let* R_α *be a primitive ring with irreducible module* V_α, *commuting ring* $D_\alpha = \mathrm{Hom}_{R_\alpha}(V_\alpha, V_\alpha)$ *and let* C_α *be the center of* D_α. *For any ultra-filter* $\mathcal{F}$ *on* $\{\alpha\}$, $R = \prod R_\alpha/\mathcal{F}$ *is a primitive ring with irreducible module* $V = \prod V_\alpha/\mathcal{F}$ *and commuting ring* $D = \prod D_\alpha/\mathcal{F}$ *with center* $C = \prod C_\alpha/\mathcal{F}$.

Proof. We prove the large part of the theorem, leaving it to the reader to show that the commuting ring is indeed as described. In what is to follow we shall not need the result about D or C.

Let $\overline{R} = \prod R_\alpha$, the complete product of the R_α . For $r \in \overline{R} = \prod R_\alpha$ we define an action on $\overline{V} = \prod V_\alpha$ by $(vr)\alpha = v(\alpha)r(\alpha)$. If $r \equiv 0\ (\mathcal{F})$ and $v \in \overline{V}$, $\{\alpha \mid (vr)(\alpha) = 0\}$ contains $\{\alpha \mid r(\alpha) = 0\}$ which is in $\mathcal{F}$ so $vr \equiv 0\ (\mathcal{F})$; similarly if $v \equiv 0\ (\mathcal{F})$ then $vr \equiv 0\ (\mathcal{F})$ for all $r \in \overline{R}$. Thus we get $R = \prod R_\alpha/\mathcal{F}$ to act on $V = \prod V_\alpha/\mathcal{F}$.

We claim that V is an irreducible R-module; for let $V \equiv 0\ (\mathcal{F})$ and let $u \in V$. Let $T_0 = \{\alpha \mid v(\alpha) \neq 0\}$; since $v \not\equiv 0\ (\mathcal{F})$, $\{\alpha \mid v(\alpha) = 0\} \notin \mathcal{F}$ hence $T_0 = A - \{\alpha \mid v(\alpha) = 0\}$ is in $\mathcal{F}$. For $\alpha \in T_0$ pick $r_\alpha \in R_\alpha$ so that $v_\alpha r_\alpha = u_\alpha$ $(= u(\alpha))$; we can do this since R_α acts primitively on V_α and $v_\alpha \neq 0$. Therefore $vr \equiv u\ (\mathcal{F})$ whence $vR = V$ confirming the irreducibility of V. Moreover R acts faithfully on V for if $vr = 0$ for all $v \in V$ and $r \in R$ if $r_1 \in \overline{R}$ is a representative for r then $\{\alpha \mid (vr_1)(\alpha) = 0\}$ is in $\mathcal{F}$ and is contained in $\{\alpha \mid r_1(\alpha) = 0\}$. Hence $\{\alpha \mid r_1(\alpha) = 0\}$ is in $\mathcal{F}$ leading to $r_1 \equiv 0\ (\mathcal{F})$ that is, $r = 0$.

We now consider prime rings satisfying a polynomial identity.

LEMMA 5.7. *If* R *is a prime ring satisfying a polynomial identity over its centroid then* R *is a left Goldie ring and a right Goldie ring.*

<u>Proof</u>. We may assume that R satisfies a multilinear identity

$$p(x_1, \ldots, x_n) = \sum_{\sigma \in S_n} \alpha_\sigma x_{\sigma(1)} \cdots x_{\sigma(n)}$$

where the α_σ are in the centroid of R and σ runs over the symmetric group S_n.

We claim that in R no direct sum of left ideals can have length n, for if $I_1 \oplus \ldots \oplus I_n$ is such a direct sum pick $x_i \in I_i$. From $0 = p(x_1, \ldots, x_n) = \Sigma\, \alpha_\sigma x_{\sigma(1)} \cdots x_{\sigma(n)}$ and from the directness of the sum we get

$$\Big(\sum_{\sigma \in S_{n-1}} \alpha_\sigma x_{\sigma(1)} \cdots x_{\sigma(n-1)}\Big) I_n = (0).$$

Since R is prime this forces

$$\sum_{\sigma \in S_{n-1}} \alpha_\sigma x_{\sigma(1)} \cdots x_{\sigma(n-1)} = 0$$

for $x_1 \in I_1, x_2 \in I_2, \ldots, x_{n-1} \in I_{n-1}$. Repeating the argument we eventually get $I_1 = (0)$, a contradiction. Similarly we get no infinite direct sums of right ideals.

We now turn to the ascending chain condition on left annihilators. Let $I_1 \subset I_2 \subset \ldots \subset I_n$ be a properly ascending chain of left annihilators where $I_i = \ell(K_i)$, $I_i \neq I_{i+1}$.

By assumption $I_i K_{i-1} \neq (0)$, $K_1 \supset K_2 \supset \ldots \supset K_n$. In $p(x_1, \ldots, x_n)$, put $x_1, \ldots, x_{n-1} \in I_n$, $x_n \in K_n$; using

$$p(x_1, \ldots, x_n) = \Sigma\, \alpha_\sigma x_{\sigma(1)} \cdots x_{\sigma(n)}$$

we get

$$K_n \sum_{\sigma \in S_{n-1}} \alpha_\sigma x_{\sigma(1)} \cdots x_{\sigma(n-1)} = 0$$

for $x_i \in I_n$. Now $(0) \neq K_{n-1} I_{n-1} \subset I_{n-1} \cap K_{n-1}$; pick $x_{n-1} \in$ $x_{n-1} \in K_{n-1} I_{n-1}$ and $x_1, \ldots, x_{n-2} \in I_{n-1}$. We get

$$K_{n-1}I_{n-1} \sum_{\sigma \in S_{n-2}} \alpha_\sigma x_{\sigma(1)} \cdots x_{\sigma(n-2)} = (0)$$

so that by the primeness of R,

$$I_{n-1} \sum_{\sigma \in S_{n-2}} \alpha_\sigma x_{\sigma 1)} \cdots x_{\sigma(n-2)} = (0)$$

for $x_1, \ldots, x_{n-2} \in I_{n-1}$. In particular, if they are in I_{n-2} except for $x_{n-2} \in K_{n-2}I_{n-2}$ we get

$$I_{n-1}K_{n-2}I_{n-2} \sum_{\sigma \in S_{n-3}} \alpha_\sigma x_{\sigma(1)} \cdots x_{\sigma(n-3)} = (0) .$$

Continuing we get $I_{n-1}K_{n-2}I_{n-2}K_{n-3} \cdots = (0)$. Since $I_iK_{i-1} \neq (0)$ we get a contradiction.

We can argue similarly for right annihilators. This proves the lemma.

We now have all the required pieces to prove an interesting theorem due to Posner.

THEOREM 5.6. Suppose that R is a prime ring satisfying a polynomial identity over its centroid. Then R can be imbedded as an order in D_n where D is a division algebra finite-dimensional over its center. Otherwise put, the ring of quotients of R exists and satisfies the same identity as does R.

Proof. Since R is a prime ring it has no non-zero nilpotent ideals. In conjunction with the fact that it is a left and right Goldie ring (Lemma 5.7) and Theorem 5.1 it has no non-zero nil ideals. Thus the polynomial ring R[t], by Theorem 1.18 of (N), is semi-simple and satisfies the identity of R (if we assume, as we may, that this identity is multilinear). In other words, we may assume that R is semi-simple. As such, by Theorem 2.39 of (N), R satisfies some standard identity $[x_1, \ldots, x_r]$. As a semi-simple ring R is a subdirect sum of primitive ring R_α, α in some index set $\mathfrak{A}$, each of which is a homomorphic image of R. As a homomorphic image of R each R_α satisfies the same standard identity as does R so is a simple ring finite-dimensional over its center.

Given $r \neq 0 \in R$ let $T_r = \{\alpha \in \mathfrak{A} \mid r(\alpha) \neq 0\}$. Since R is a prime ring, given $a \neq 0$, $b \neq 0$ in R there is an $x \in R$ such that $axb \neq 0$; but then $T_{axb} \neq 0 \subset T_a \cap T_b$. Therefore we can enlarge $\{T_r \mid r \neq 0 \in R\}$ to a filter and so to an ultra-filter $\mathcal{F}$ on $\mathfrak{A}$. From our construction we have that R is isomorphically imbedded in $S = \prod_{\alpha \in} R_\alpha/\mathcal{F}$. By Theorem 5.5, S is a primitive ring and we already know it satisfies $[x_1, \ldots, x_r]$ since each R_α does, hence S is a simple algebra finite-dimensional over its center Z. Consider $RZ \subset S$.

Any identity over the prime field P of Z satisfied by RZ is satisfied by R (if $P = GF(p)$, $RP \subset R$; if P is the rational field by clearing of denominators we get an identity with integer coefficients which then holds in R) hence in each R_α and so in $S = \prod R_\alpha/\mathcal{F}$. By Theorem 5.4, $RZ = S$.

Given $r \in R$ which is not a zero divisor in R it is not a zero divisor in RZ. If it were, pick $w \neq 0 \in RZ$ such that $w = s_1\lambda_1 + \ldots + s_k\lambda_k$, $s_i \neq 0 \in R$, $\lambda_i = 0 \in Z$ of shortest length $(= k)$ such that $rw = 0$. Now $r(wxrs_1 - s_1xrw) = 0$ for any $x \in R$; but $wxrs_1 - s_1xrw = (s_2xrs_1 - s_1xrs_2)\lambda_2 + \ldots + (s_kxrs_1 - x_1xrs_k)\lambda_k$ is of shorter length than w and annihilates r, hence $wxrs_1 - s_1xrw = 0$, that is $wxrs_1 = 0$ for all $x \in R$. Hence $wRZrs_1 = (0)$; since RZ is simple (being S) and $w \neq 0$ we conclude that $rs_1 = 0$, contrary to the assumption that r is not a zero divisor in R. Now as a non-zero divisor in a finite-dimensional algebra r must be invertible in RZ. We have shown that all regular elements in R are invertible in RZ.

By Lemma 5.7, R is a prime Goldie ring so has a ring of quotients $Q(R) = D_n$, D a division ring. We can imbed Q(R) in RZ, for every $x \in Q(R)$ is of the form $x = a^{-1}b$, $a, b \in R$, a regular; map it on $a^{-1}b$ in RZ, where a^{-1} is the inverse of a in RZ. Thus Q(R) satisfies the identities over P that R does, hence D also does. By Theorem 2.38 of (N), D is finite-dimensional over its center. The theorem is now completely proved.

Bibliography

1. I. N. Herstein and Lance Small, "Nil rings satisfying certain chain conditions", Canadian Journal, 16 (1964), pp. 771-776.

2. C. Lanski, "Nil subrings of Goldie rings", (to appear).

3. E. Posner, "Prime rings satisfying a polynomial identity", Proc. Amer. Math. Soc., 11 (1960), pp. 180-184.

4. Lance Small, Thesis , University of Chicago, 1965.

CHAPTER 6

DECOMPOSITION IN NOETHERIAN RINGS[1]

The topic treated in this chapter is an attempt to simulate in a non-commutative Noetherian ring the Lasker-Noether primary decomposition in a commutative Noetherian ring. For a large part, the results we shall give are due to Lesieur and Croisot.

A decomposition is given for an arbitrary left ideal as an intersection of special left ideals -- the so-called tertiary ones -- and one could hope that this would be a suitable vehicle for a profound study of non-commutative Noetherian rings. Unfortunately, at least as yet, this has not been the case; we know of no effective application of these ideas to obtain the central, deep and not-yet-established theorems that would reveal the make up of a general Noetherian ring.

Let us consider what might be considered a down-to-earth question, whose answer would generalize a known theorem of Levitzki: Let R be a left Noetherian ring and λ, μ left ideals of R; suppose that for all $a \in \lambda$, $a^{n(a)} \in \mu$ for a suitable integer $n(a)$; is it then true that $\lambda^k \subset \mu$ for some integer k? This would seem to be a natural target for such a decomposition theory yet in this general setting the theory falls short of providing an answer. However, in all fairness, making use of some of its concepts if not the results themselves we have been able to establish the special case of the result in which $a^n \in \mu$ for all $a \in \lambda$,

---

[1] I am indebted to A. Buccino for his great help in organizing and collating this material.

n a fixed integer, and in the general case when R satisfies a polynomial identity.

All rings R to be considered will be left-Noetherian, that is, satisfy the ascending chain condition on left ideals, and will have unit elements. All modules M will be unital, finitely generated left R-modules.

DEFINITION. An element $a \in R$ is an annihilating element for M if $aN = (0)$ for some submodule $N \neq (0)$ of M. A left ideal I of R is an annihilating left ideal for M if $IN = (0)$ for some submodule $N \neq (0)$ of M.

We denote by $O(M)$ the ideal $\{a \in R \mid aM = (0)\}$ of R.

The next few definitions make sense for any ring and all or parts of the very early lemmas hold in this more general context, but all these will be of most interest in the Noetherian case.

DEFINITION. If N is a submodule of M then the primary radical of N, written rad(N), is the intersection of all prime ideals of R which contain $O(M/N)$. N is a primary submodule of M if all annihilating elements for M/N are in rad(N).

A remark which holds for any ring R: if $x \in \text{rad}(N)$ then $x^t \in O(M/N)$ for some t depending on x. This is true since the intersection of all prime ideals of a ring is a nil ideal.

One further definition at this point:

DEFINITION. The tertiary radical of N, N a submodule of M, written t-rad(N), is the set of all elements of R which are annihilating elements for all non-trivial submodules of M/N.

One immediately sees that:

$$\text{t-rad}(N) = \{a \in R \mid \text{for all } \alpha \notin N \text{ there exists } \beta \in M, \beta \notin N \text{ with } aR\beta \subset N\}.$$

LEMMA 6.1 *Let* N *be a submodule of* M; *if* $a_1, \ldots, a_n \in$ t-rad(N) *then, given* $\alpha \in M$, $\alpha \notin N$ *there is a* $\beta \in R\alpha$, $\beta \notin N$ *such that* $a_i R\beta \subset N$ *for* $i = 1, 2, \ldots, n$.

Proof. We go by induction on n. If $n = 1$ this is merely the definition of t-rad(N). Suppose then that we have found a $\gamma \in R\alpha$, $\gamma \notin N$ such that $a_i R\gamma \subset N$ for $i = 1, 2, \ldots, n-1$. Since $a_n \in$ t-rad(N) there is a $\beta \in R\gamma \subset R\alpha \beta \notin N$ such that $a_n R\beta \subset N$. However for $i < n$, $a_i R\beta \subset a_i R\gamma \subset N$, thereby the lemma is proved.

COROLLARY 1. t-rad(N) *is a two-sided ideal of* R.

Proof. From the very definition of t-rad(N) we immediately have that $aR \subset$ t-rad(N) and $Ra \subset$ t-rad(N) for all $a \in$ t-rad(N). To finish we merely need that $a, b \in$ t-rad(N) forces $a-b \in$ t-rad(N); this is however clear from the lemma.

COROLLARY 2. *Let* I = t-rad(N); *given* $\alpha \notin N$ *there is a* $\beta \in R\alpha$ *such that* $\beta \notin N$, $I\beta \subset N$.

Proof. Since R is left Noetherian and I is an ideal of R, $I = Ra_1 + \ldots + Ra_n$ for appropriate $a_i \in I$. Pick by the lemma $\beta \in R\alpha$, $\beta \notin N$ such that $a_i\beta \in N$ for $i = 1, \ldots, n$; then $I\beta \subseteq N$.

COROLLARY 3. *Let* I = t-rad(N) *and let* $N' = \{\alpha \in M \mid I\alpha \subset N\}$. *Then* $N' \supset N$ *and* $N' \neq N$.

Proof. Since I is a 2-sided ideal of R, N' is a submodule of M. Clearly, $N' \supset N$. By Corollary 2, we can find $\beta \notin N$ such that $I\beta \subset N$; since $\beta \in N'$, $\beta \notin N$ this finishes the proof.

In fact, we have proved a good deal more about the nature of N' for we know that in each $R\alpha$, $\alpha \notin N$ there are elements of N'. We formalize this in Lemma 6.2 but first a definition.

As usual, and as used before in these notes, M is said to be an <u>essential extension</u> of A modulo B, $B \subset A$ submodules of M, if whenever C is a submodule of M such that $C \cap A \subset B$ then $C \subset B$. That is, in M/B the module A/B meets all non-zero submodules non-trivially.

LEMMA 6.2. M *is an essential extension of* N' *modulo* N.

Proof. This is immediate from Corollary 2 to Lemma 6.1.

Before proceeding, we examine the relation of $\mathrm{rad}(N)$ to $t\text{-}\mathrm{rad}(N)$.

LEMMA 6.3. $\mathrm{rad}(N) \subset t\text{-}\mathrm{rad}(N)$.

Proof. As we remarked earlier, for all $x \in \mathrm{rad}(N)$, $x^{n(x)} \in t\text{-}\mathrm{rad}(N)$. Since R is Noetherian by a result of Levitzki (or using results of Chapter 5) $(\mathrm{rad}(N))^k \subset t\text{-}\mathrm{rad}(N)$. From the definition of $t\text{-}\mathrm{rad}(N)$ this implies that $\mathrm{rad}(N) \subset t\text{-}\mathrm{rad}(N)$.

In the commutative case we can now easily establish equality for these two radicals.

THEOREM 6.1. *If* R *is a commutative Noetherian ring and if* N *is a submodule of* M *then* $\mathrm{rad}(N) = t\text{-}\mathrm{rad}(N)$.

Proof. Lemma 6.3 already tells us that $\mathrm{rad}(N) \subset t\text{-}\mathrm{rad}(N)$. Now for the other direction.

Since R is Noetherian and M is a finitely generated unital R-module, M has the ascending chain condition on submodules. Let $a \in t\text{-}\mathrm{rad}(N)$ and let $J_n = \{\alpha \in M \mid Ra^n\alpha \subset N\}$. Then J_n form an ascending chain of submodules of M hence for some integer n, $J_n = J_{n+1}$. If $a^nM \not\subset N$ we can find $\alpha \in M$ with $a^n\alpha \notin N$; since $a \in t\text{-}\mathrm{rad}(N)$ there is an $r \in R$ with $ra^n\alpha \notin N$ but for which $aRra^n\alpha \subset N$. Since R is commutative this yields $Ra^{n+1}r\alpha \subset N$, that is, $r\alpha \in J_{n+1} = J_n$ hence $a^nr\alpha \in N$, contrary to $a^nr\alpha \notin N$.

Thus $a \in$ t-rad(N) implies $a^k \in$ rad(N). Since rad(N) is the intersection of prime ideals this forces $a \in$ rad(N), that is, t-rad(N) $\subset$ rad(N). This proves the theorem.

DEFINITION. N is a *tertiary submodule* of M if the annihilating elements for M/N are all in t-rad(N).

A quick verification reveals that N is a tertiary submodule of M if and only if t-rad(N) = $\{a \in R \mid$ there is a $\beta \notin N$ with $aR\beta \subset N\}$.

LEMMA 6.4. *Let* N *be a tertiary submodule of* M; *then* P = t-rad(N) *is a prime ideal of* R.

Proof. By Lemma 6.2, since $N' \neq N$, there exists a $\beta \in M$ $\beta \notin N$ such that for all $a \in P$, $aR\beta \subset N$. Since N is tertiary P consists of *all* elements $a \in R$ such that $aR\beta \subset N$.

Suppose $x, y \in R$ are such that $xRy \subset P$; then $xRyR\beta \subset N$. If $y \notin P$ then $yR\beta \not\subset N$ so by the tertiary nature of N we get $x \in P$. Therefore $xRy \subset P$ implies that $x \in P$ or $y \in P$, hence P is a prime ideal of R.

If N is a tertiary submodule of M and the prime ideal P of R is t-rad(N) then we say that N is *P-tertiary* and that P is the *prime ideal of N*.

A submodule N of M is called *irreducible* if it is not the intersection of two strictly larger submodules.

LEMMA 6.5. *If* N *is an irreducible submodule of* M *then it is tertiary.*

Proof. If N is not tertiary there exist $a \notin$ t-rad(N), $\alpha \notin N$ such that $aR\alpha \subset N$. Since $a \notin$ t-rad(N) there is a $\gamma \notin N$ with $aR\gamma' \subset N$ where $\gamma' \in R\gamma$ implies that $\gamma' \in N$.

Let $N^* = (N + R\alpha) \cap (N + R\gamma)$; clearly $N^* \supset N$. For $\beta \in N^*$, $\beta = \alpha' + r\alpha = \alpha'' + s\gamma$ with $\alpha', \alpha'' \in N$. Thus $s\gamma = (\alpha' - \alpha'') + r\alpha$ so $aRs\gamma \subset aR\alpha \subset N$. Since $\gamma' = s\gamma \in R\gamma$ satisfies $aR\gamma' \subset N$ we have that $\gamma' \in N$, hence $\beta \in N$. Thus $N^* \subset N$. We have exhibited N as an intersection of larger submodules. This proves the lemma.

DEFINITION. The decomposition $N = N_1 \cap \ldots \cap N_r$ of N by the N_i is <u>irredundant</u> if no N_i can be omitted.

LEMMA 6.6. <u>If</u> $N = N_1 \cap \ldots \cap N_r$ <u>is an irredundant decomposition of</u> N <u>by</u> P_i<u>-tertiary submodules</u> N_i, $i = 1, 2, \ldots, r$, <u>then</u> $\text{t-rad}(N) = P_1 \cap \ldots \cap P_r$.

<u>Proof.</u> Let $a \in P_1 \cap \ldots \cap P_r$ and let $\alpha \in M$, $\alpha \notin N$. Since $\alpha \notin N$, α is not in some N_i, say $\alpha \notin N_1$. From the fact that $a \in \text{t-rad}(N_1)$ there is an $\alpha_1 \in R\alpha$, $\alpha_1 \notin N_1$ such that $aR\alpha_1 \subset N_1$. If $\alpha_1 \in N_2$ then $aR\alpha_1 \subset N_1 \cap N_2$; if $\alpha_1 \notin N_2$ there is an element $\alpha_2 \in R\alpha_1 \subset R\alpha$, $\alpha_2 \notin N_2$ such that $aR\alpha_2 \subset N_2$. Since $aR\alpha_2 \subset aR\alpha_1 \subset N_1$ we have $aR\alpha_2 \subset N_1 \cap N_2$. Continuing this way we get an element $\beta \in R\alpha$, $\beta \notin N$ such that $aR\beta \subset N_1 \cap \ldots \cap N_r = N$. Thus by the definition of t-rad(N), $a \in \text{t-rad}(N)$ hence $P_1 \cap \ldots \cap P_r \subset \text{t-rad}(N)$.

Suppose now that $a \in \text{t-rad}(N)$ and that $\alpha \in N_j$ for all $j \neq i$, $\alpha \notin N_i$. Since N is the irredundant intersection of the N_i, $\alpha \notin N$. Therefore there is a $\beta \in R\alpha$, $\beta \notin N$ such that $aR\beta \subset N \subset N_j$. Since $\beta \in N_j$ for $j \neq i$ and $\beta \notin N$ we must have that $\beta \notin N_i$. Since N_i is P_i-tertiary we conclude that $a \in P_i$. But then $a \in \bigcap_{i=1}^{r} P_i$, whence $\text{t-rad}(N) \subset P_1 \cap \ldots \cap P_r$.

LEMMA 6.7. <u>If</u> N_1 <u>and</u> N_2 <u>are</u> P-<u>tertiary submodules of</u> M <u>then so is</u> $N_1 \cap N_2$.

<u>Proof</u>. Let $N = N_1 \cap N_2$; by Lemma 6.5, t-rad(N) = P. To finish the proof we merely must show that N is tertiary.

Let $a \in R$, $\beta \in M$, $\beta \notin N$ such that $aR\beta \subset N$. Since $\beta \notin N$, $\beta \notin N_1$ or $\beta \notin N_2$. If $\beta_1 \notin N_1$, since N_1 is tertiary, $a \in$ t-rad$(N_1) = P$; similarly, if $\beta \notin N_2$, $a \in P$. Thus $aR\beta \subset N$, $\beta \notin N$ implies that $a \in P =$ t-rad(N). Thus N is tertiary.

DEFINITION. A decomposition $N = N_1 \cap \ldots \cap N_r$ of a submodule N by P_i-tertiary submodules N_i is called <u>reduced</u> if it is irredundant and the P_i are distinct.

We now can prove the analog of the primary decomposition for commutative Noetherian rings.

THEOREM 6.2. <u>Every submodule</u> N <u>of</u> M <u>has a reduced decomposition</u> $N = N_1 \cap \ldots \cap N_r$ <u>where</u> N_i <u>is</u> P_i<u>-tertiary</u>.

<u>Proof</u>. That every submodule has an intersection of a finite number of irreducible ones is easy, just as in the commutative case. Merely consider the set of submodules which have no such representation; if this set is non-empty it has a maximal element. However, this maximal element is therefore irreducible, giving a contradiction.

The rest follows from Lemmas 6.4, 6.5, 6.6, and 6.7.

We now want to establish the uniqueness of the associated primes. To this end we prove

LEMMA 6.8. <u>If</u> N, U, V, U', V' <u>are submodules of</u> M <u>such that</u> U <u>is</u> P-<u>tertiary</u>, U' <u>is</u> P'-<u>tertiary</u>, $P \neq P'$ <u>and</u> $N = U \cap V = U' \cap V'$ <u>then</u> $N = V \cap V'$.

<u>Proof</u>. Let $\alpha \in V \cap V'$. Since $P \neq P'$, there is an element in one and not in the other; say $a \in P$, $a \notin P'$. If $\alpha \in N$ there exists a $\beta \in R\alpha$, $\beta \notin N$ such that $aR\beta \subset N$. Thus $aR\beta \subset U' \cap V'$.

Since $\beta \notin N$, $\beta \notin U'$. Since U' is P'-tertiary and $aR\beta \subset U'$ we conclude that $a \in P'$, a contradiction. Thus $\alpha \in N$. That is, $V \cap V' \subset N$; since clearly $N \subset V$, $N \subset V'$ we get $N \subset V \cap V'$. This proves the lemma.

The lemma immediately yields

THEOREM 6.3. *If the submodule* N *of* M *has the two reduced decompositions*

$$N = N_1 \cap \ldots \cap N_r = N_1' \cap \ldots \cap N_s'$$

then $r = s$ *and the set of prime ideals* P_i *of the* N_i *coincides with the set of prime ideals* P_i' *of the* N_i'.

Proof. We show $P_i' = P_i$ for some i. Suppose not. Since $P_1' = P_i$, by Lemma 6.8, since $N = N_1 \cap \ldots \cap N_r = N_1' \cap \ldots \cap N_s'$, $N = N_2 \cap \ldots \cap N_r \cap N_2' \cap \ldots \cap N_s'$. Since $P_i' \neq P_2$ we get $N = N_3 \cap \ldots \cap N_r \cap N_2' \cap \ldots \cap N_s'$. Continuing we arrive at $N = N_2' \cap \ldots \cap N_s'$ contrary to the irredundancy of the representation $N = N_1' \cap \ldots \cap N_s'$. Thus $P_i' = P$.

In the same way given j, $P_j' = P_k$ for some k. This shows $s \leq r$. The argument is symmetric, $r \leq s$, and $\{P_i'\} = \{P_i\}$.

The theory developed here does not have many of the desirable properties of the primary decomposition theory for commutative Noetherian rings. For instance, the tertiary radical of a left ideal I of R, considered as an R-module does not necessarily contain I. An even greater shortcoming is that one can find left Noetherian rings R and a module M with submodules $N_1 \subset N_2$ for which $\text{t-rad}(N_1) \not\subset \text{t-rad}(N_2)$. In fact one can show that the t-radical of M is monotonic if and only if for all submodules N of M, $\text{rad}(N) = \text{t-rad}(N)$, and so if all tertiary modules are primary.

Emmy Noether gave a simple example of a situation where tertiary modules are not always primary. Let Q be the field of of rational numbers and let R be the algebra generated over Q by a, b, c where $a^2 = a$, $b^2 = b$, $c^2 = ab = ba = ca = bc = 0$ and $ac = cb = c$. It is easy to verify that in the R-module R, the submodule (left ideal) Ra is tertiary but not primary.

We now examine when tertiary modules are primary.

DEFINITION. An R-module is called an <u>Artin-Reese module</u> if for each left ideal I of R and each submodule N of M and each integer $n \geq 0$ we can find an integer $h = h(n) \geq 0$ such that $I^h M \cap N \subset I^n N$.

LEMMA 6.9. <u>A</u> P-<u>tertiary submodule</u> N <u>of</u> M <u>is primary if and only if</u> $P^n M \subset N$ <u>for some</u> n.

<u>Proof</u>. If N is primary then rad(N) = t-rad(N) = P and so since $x \in$ rad(N) implies $x^{k(x)} \in O(M/N)$, by a theorem of Levitski, $P^n \subset O(M/N)$ (see Lemma 6.3). Hence $P^n M \subset N$.

On the other hand, if $P^n M \subset N$ then $P^n \subset$ rad(N); since rad(N) is an intersection of prime ideals this puts P in rad(N). Together with rad(N) $\subset$ P we get the result.

THEOREM 6.4. <u>An</u> R-<u>module</u> M <u>is an Artin-Reese module if and only if every tertiary submodule of</u> M <u>is primary</u>.

<u>Proof</u>. Suppose that M is an Artin-Reese module and that N is a P-tertiary submodule of M. Let $N' = \{ \alpha \in M | P\alpha \subset N \}$, then there exists an integer n so that $P^n M \cap N' \subset PN' \subset N$. Thus by Lemma 6.2, $P^n M \subset N$; Lemma 6.9 then tells us that N is primary.

Suppose on the other hand, that all tertiary submodules of M are primary. Let N be a submodule of M and n an integer, and I a left ideal of R. Let $I^n N = N_1 \cap \ldots \cap N_r \cap N'_1 \cap \ldots \cap N'_s$ be a reduced tertiary decomposition of $I^n N$ where each N_i is P_i-tertiary and each N'_j is P'_j-tertiary and where $I^n \subset P_i$ for

$i = 1, \ldots, r$ and $I^n \not\subset P'_j$ for $j = 1, 2, \ldots, s$. Since tertiary modules are primary there exists an integer h such that $I^h M \subset N_1 \cap \ldots \cap N_r$. Now $N \subset N'_1 \cap \ldots \cap N'_s$ for if $\alpha \in N$, $a \in I^n$ then $aR\alpha \subset I^n N \subset N'_1 \cap \ldots \cap N'_s$. If $\alpha \notin N_1 \cap \ldots \cap N'_s$ then $a \in P'_j$ for some j since the N'_j are tertiary. This would be a contradiction. Thus $I^h M \cap N \subset I^n N$.

COROLLARY (Artin-Reese): Every finitely generated module over a commutative Noetherian ring is an Artin-Reese module.

Proof. This is merely a combination of Theorems 6.1 and 6.4.

Bibliography.

1. I. N. Herstein, "A theorem on left Noetherian rings", Jour. Math. Anal. Appl, 15 (1966), pp. 91-96.

2. L. Lesieur and R. Croisot, "Sur la decomposition en ideaux primaires dans un anneau non necessairment commutatif", Comptes Rendus Acad. Sci. Paris 243 (1955), pp. 1988-1991.

3. L. Lesieur and R. Croisot, "Theorie Noetherienne des anneaux, des demi-groupes et des modules dans le cas non commutatif I, II, III", in Colloque Algebre Superieure Bruxelles 1956, Math. Annalen 134 (1958), pp. 485-496, and Acad. Roy. Belg. Bull. Cl. Sci. (5), 44(1958), pp. 75-93 respectively.

4. J. A. Riley, "Axiomatic primary and tertiary decomposition theory", Trans. Amer. Math. Soc. 105 (1962), pp. 177-201.

CHAPTER 7

RECENT EXAMPLES AND COUNTER-EXAMPLES

In a remarkable paper published recently, Golod and Shafarevitch proved an extremely powerful theorem -- although rather easy to prove -- which has provided a technique to tackle a wide range of problems. As an immediate consequence of their work one can construct a nil, but not locally nilpotent, algebra thereby providing a negative answer to the Kurosh Problem; one can construct a torsion group which is not locally finite, thereby settling in the negative the general Burnside Problem; one can construct an infinite class field tower, in fact, the field $Q(\sqrt{-3 \cdot 5 \cdot 7 \cdot 11 \cdot 13 \cdot 19})$ carries such a tower above it. It would seem that these are just the beginning of the applications of their theorem.

What is perhaps even more remarkable about what they have done is the beautiful simplicity of their approach. Part of this chapter is devoted to an exposition of a portion of their work; we shall develop it enough to settle the Kurosh and Burnside Problems. We follow the treatment used by Jon Alperin in the Algebra Seminar at the University of Chicago in April, 1964.

Let F be any field and let $T = F[x_1, \ldots, x_d]$ be the polynomial ring over F in the d non-commuting variables $x_1, x_2, \ldots, x_d$. T has a degree function and carries the structure of a graded algebra; T can be written as

$$T = T_o \oplus T_1 \oplus \ldots \oplus T_n \oplus \ldots$$

where $T_o = F$ and where T_n has a basis of d^n elements consisting of the monomials $x_{i_1} x_{i_2} \ldots x_{i_n}$. The elements of T_n are said to be homogeneous of degree n.

Let $\mathfrak{a} = (f_1, f_2, \ldots)$ be the ideal of T generated by the homogeneous elements $f_1, f_2, \ldots$ of degrees $2 \leq n_1 \leq n_2 \leq \ldots$ respectively. Let r_i be the number of n_j which are equal to i.

Now the algebra $A = T/\mathfrak{a}$ inherits the grading of T, since $\mathfrak{a}$ is homogeneously generated, in fact $A = A_0 \oplus A_1 \oplus \ldots \oplus A_n \oplus \ldots$ where $A_n \approx T/(\mathfrak{a} \cap T_n)$. Let $b_n = \dim_F(A_n)$.

The Golod-Shafarevitch theorem gives a sufficient condition that A be infinite dimensional over F. This is the ultra-important

THEOREM 7.1. (Golod-Shafarevitch) <u>For A as described above</u>

1. $b_n \geq db_{n-1} - \sum_{n_i \leq n} b_{n-n_i}$ <u>for</u> $n \geq 1$.

2. <u>if</u> $r_i \leq (\frac{d-1}{2})^2$ <u>then</u> A <u>is infinite-dimensional over</u> F.

<u>Proof</u>. We will exhibit linear mappings φ, ψ so that the sequence

$$(1) \quad A_{n-n_1} \oplus \ldots \oplus A_{n-n_k} \oplus \ldots \xrightarrow{\varphi} \overbrace{A_{n-1} \oplus \ldots \oplus A_{n-1}}^{d\text{-times}} \xrightarrow{\psi} A_n \to 0$$

where the first sum runs over all $n_i \leq n$, is exact. Note that if we could do this then (1) would be proved for then $db_{n-1} = b_n + \dim \operatorname{Ker} \psi$. Since $\ker \psi$ is a homomorphic image of $A_{n-n_1} \oplus \ldots \oplus A_{n-n_k} \oplus \ldots$, $\dim \operatorname{Ker} \psi \leq \sum_{n_i \leq n} b_{n-n_i}$, hence $db_{n-1} \leq b_n + \sum_{n_i \leq n} b_{n-n_i}$, the required result.

Our task has been reduced to defining the φ and ψ. To do this we shall define mappings Φ and Ψ for the sequence

$$(2) \quad T_{n-n_1} \oplus \ldots \oplus T_{n-n_k} \oplus \ldots \xrightarrow{\Phi} T_{n-1} \oplus \ldots \oplus T_{n-1} \xrightarrow{\Psi} T_n \to 0$$

where Ψ and Φ are linear and the sequence is exact at T_n. From this we shall get the φ, ψ for the sequence (1).

We define Ψ by:

$$\Psi : t_1 \oplus \ldots \oplus t_d \longrightarrow \sum_{i=1}^{d} t_i x_i .$$

To get Φ, map $s_{n-n_1} \oplus s_{n-n_2} \oplus \ldots \to \sum s_{n-n_i} f_i$; since $\sum_{n_i \leq n} s_{n-n_i} f_i = \sum_{i=1}^{d} u_i x_i$ with $u_i \in T_{n-1}$ define

$$\Phi : S_{n-n_1} \oplus S_{n-n_2} \oplus \ldots \to u_1 \oplus u_2 \oplus \ldots \oplus u_d .$$

By the uniqueness of representation of any element in T_n as $\sum_{i=1}^{d} u_i x_i$, $u_i \in T_{n-1}$, Φ is well-defined. Trivially Φ and Ψ are linear. It is equally clear that Ψ is onto T_n so the sequence (2) is exact at T_n.

Let $\mathfrak{a}_i = \mathfrak{a} \cap T_i$; we show that we can induce (1) from (2). If $t_1, \ldots, t_d \in \mathfrak{a}_{n-1}$ then clearly $\sum t_i x_i \in \mathfrak{a}$ and $\sum t_i x_i \in T_n$ hence $\sum t_i x_i \in \mathfrak{a}_n$, so Ψ induces down to

$$\psi : A_{n-1} \oplus \ldots \oplus A_{n-1} \xrightarrow{\psi} A_n \longrightarrow 0.$$

We consider Φ. Suppose that $s_{n-n_1}, s_{n-n_2}, \ldots, s_{n-n_i}$ are in $\mathfrak{a}_{n-n_1}, \mathfrak{a}_{n-n_2}, \ldots, \mathfrak{a}_{n-n_i}$ respectively. We want to show that $u_1, \ldots, u_d$ defined by $\sum s_{n-n_i} f_i = \sum u_i x_i$ are in

$\mathfrak{a}_{n-1}$. Since Φ is linear it is enough to do it for $s_{n-n_i} \in \mathfrak{a}_{n-n_i}$. Now $s_{n-n_i} f_i = \sum_{j=1}^{d} s_{n-n_i} g_{ij} x_j$ where $f_i = \sum g_{ij} x_j$; if $s_{n-n_i} \in \mathfrak{a}$ then $s_{n-n_i} g_{ij} \in \mathfrak{a}$. But $s_{n-n_i} g_{ij} = u_j$ hence $u_j \in \mathfrak{a} \cap T_{n-1} = \mathfrak{a}_{n-1}$. Thus Φ induces down.

All that remains is to show exactness at $A_{n-1} \oplus \ldots \oplus A_{n-1}$. We first show that $\varphi\psi = 0$. If $s_{n-n_1}, s_{n-n_2}, \ldots \in T_{n-n_1}$, $T_{n-n_2}, \ldots, (s_{n-n_1} \oplus s_{n-n_2} \oplus \ldots)\Phi\Psi = \sum_{i=1}^{d} u_i x_i$ where $\sum u_i x_i = \sum s_{n-n_i} f_i$; since the $f_i \in \mathfrak{a}$, $\sum s_{n-n_i} f_i \in \mathfrak{a}$ hence $\sum u_i x_i \in \mathfrak{a}$ whence $\Phi\Psi$ maps $T_{n-n_1} \oplus \ldots$ into $\mathfrak{a}$ and so $A_{n-n_1} \oplus A_{n-n_2} \oplus \ldots$ is mapped into 0 by $\varphi\psi$. We now want to show that if $(t_1 \oplus \ldots \oplus t_d)\Psi \in \mathfrak{a}$ then we can find $u_1, u_2, \ldots, u_d \in T_{n-1}$ such that $t_i - u_i \in \mathfrak{a}$ for $i = 1, 2, \ldots, d$ and such that $\sum u_i x_i = \sum s_{n-n_i} f_i$ for some s_{n-n_i} in the appropriate T_{n-n_i}.

Suppose then that $\sum t_i x_i \in \mathfrak{a}$; being in $\mathfrak{a}$ which is generated as an ideal by the f_j, $\sum t_i x_i = \sum a_{k\ell} f_\ell b_{k\ell} + \sum c_\ell f_\ell$ where the $a_{k\ell}$, $b_{k\ell}$, c_ℓ are homogeneous and $b_{k\ell}$ is of degree ≥ 1. Since $\sum t_i x_i \in T_n$ by picking out the parts of the right degree we may assume that the $a_{k\ell} f_\ell b_{k\ell}$, $c_\ell f_\ell$ are all in T_n. Since the $b_{k\ell}$ are of degree at least 1, $b_{k\ell} = \sum d_{k\ell m} x_m$ hence

$$\sum a_{k\ell} f_\ell b_{k\ell} = \sum a_{k\ell} f_\ell d_{k\ell m} x_m = \sum d_m x_m ,$$

where $d_m = \sum_{k,\ell} a_{k\ell} f_\ell d_{k\ell m}$ so is in $\mathfrak{A}$. If $\sum c_\ell f_\ell = \sum_{i=1}^{d} u_i x_i$ we have $\sum_{i=1}^{d} t_i x_i = \sum_{m=1}^{d} d_m x_m + \sum_{i=1}^{d} u_i x_i$, hence $t_i - u_i = d_i \in \mathfrak{A}$. But $(c_1 \oplus \ldots \oplus c_i \oplus \ldots)\Phi = u_1 \oplus \ldots \oplus u_d$ by definition of Φ. Hence we have proved the exactness at $A_{n-1} \oplus \ldots \oplus A_{n-1}$. This proves part (1) of the theorem.

We now consider formal power series in t with integer coefficients; we declare $\sum_{n=1}^{\infty} c_n t^n \geq \sum_{n=1}^{\infty} d_n t^n$ if $c_i \geq d_i$ for all i. From part (1) we have

$$\sum_{n=1}^{\infty} b_n t^n \geq \sum_{n=1}^{\infty} db_{n-1} t^n - \sum_{n=1}^{\infty} \sum_{n_i \leq n} b_{n-n_i} t^n .$$

From the definition of r_i,

$$\sum_{n_i, m} b_m t^{n_i + m} = \sum t^{n_i} \sum_{m=0}^{\infty} b_m t^m = \left(\sum_{i=2}^{\infty} r_i t^i\right)\left(\sum_{m=0}^{\infty} b_m t^m\right).$$

Let $P_A(t) = \sum_{m=0}^{\infty} b_m t^m$. From the above

$$P_A(t) - 1 \geq dtP_A(t) - \left(\sum_{i=2}^{\infty} r_i t^i\right) P_A(t),$$

hence

$$P_A(t)\left(1 - dt - \sum_{i=2}^{\infty} r_i t^i\right) \geq 1 .$$

But then we immediately see that if the coefficients of $(1 - dt + \sum_{i=2}^{\infty} r_i t^i)^{-1}$ are positive then an infinite number of the b_n are not 0, hence A is infinite dimensional.

THEOREM 7.2. A <u>is infinite dimensional over</u> F <u>if the coefficients in the power series expansion of</u>

$$(1 - dt + \sum_{i=2}^{\infty} r_i t^i)^{-1}$$ <u>are positive.</u>

We leave it as an excercise to show that if $r_i \leq (\frac{d-1}{2})^2$ for all i then indeed all the coefficients of

$$(1 - dt + \sum_{i=2}^{\infty} r_i t^i)^{-1}$$ are positive. Thus part (2) of Theorem 7.1 is proved.

We use Theorem 7.1 to construct a finitely generated nil algebra which is not nilpotent. This settles the Kurosh Problem in the negative.

THEOREM 7.3. <u>If</u> F <u>is any countable field there exists an infinite dimensional nil algebra over</u> F <u>generated by three elements.</u>

<u>Proof.</u> Let $T = F[x_1, x_2, x_3]$ as before be the free algebra over F generated by the three non-commuting indeterminates x_1, x_2, x_3; T is graded, $T = T_o \oplus T_1 \oplus \ldots \oplus T_n \oplus \ldots$. Consider the ideal $T' = T_1 \oplus T_2 \oplus \ldots$ of T. T' is countable so we enumerate its elements as $s_1, s_2, \ldots$. Let $m_1 \geq 2$ be an integer and let $s_1^{m_1} = s_{12} + s_{13} + \ldots + s_{1k_1}$ where $s_{1j} \in T_j$. Choose an integer $m_2 > 0$ so that $s_2^{m_2} \in T_{k_1+1} \oplus \ldots$; $s_2^{m_2} = s_{2,k_1+1} + \ldots + s_{2,k_2}$ with $s_{2j} \in T_j$. Thus we may choose integers $k_1 < k_2 < \ldots$ such that $s_i^{m_i} = s_{i,k_{i-1}+1} + \ldots + s_{ik_i}$ with $s_{ij} \in T_j$. Let $\mathfrak{a}$ be the ideal of T generated by the s_{ij}. By construction of the s_{ij}, r_k in Theorem 7.1 is at most 1. Since d = 3, we have $r_i \leq (\frac{d-1}{2})^2$. By Theorem 7.1, $T/\mathfrak{a}$ is infinite dimensional,

over F. Moreover, by construction $T'/\mathfrak{a}$ is a nil algebra. This is the required example since $T'/\mathfrak{a}$ is generated over F by three elements.

This result is not the best possible for as Golod has shown (in a proof which is not difficult) *given any field* F *there is an infinite dimensional nil algebra over* F *generated by two elements.*

We use Theorem 7.3 to construct an example of an infinite torsion group generated by three elements, which is therefore a counter-example to the Burnside Problem.

THEOREM 7.4. *If* p *is any prime number there is an infinite group* G *generated by three elements in which every element has order a power of* p.

Proof. Let F be the prime field with p elements and let $\mathfrak{a}$ be the ideal in $T = F[x_1, x_2, x_3]$ constructed during the proof of Theorem 7.3. Let $A = T/\mathfrak{a}$ and let a_1, a_2, a_3 be $x_1 + \mathfrak{a}$, $x_2 + \mathfrak{a}$, $x_3 + \mathfrak{a}$ respectively. Let G be the multiplicative semi-group generated by $1 + a_1$, $1 + a_2$ and $1 + a_3$. If any element is in G it is of the form $1 + a$ where $a \in T'/\mathfrak{a}$ hence a is nilpotent; if $a^n = 0$ then $a^{p^n} = 0$ so that $(1+a)^{p^n} = 1$ since we are in characteristic p. Thus G is a group in which every element has order a power of p. We claim G is infinite, for if G is finite the linear combinations of elements in G over F would form a finite dimensional subalgebra B of A. But since $1 + a_i$, 1 are in G, $a_i \in B$ hence $A = B$, contradicting that A is infinite dimensional. This finishes the proof.

We now turn to other things with the advice to the reader to read the entire Golod-Shafarevitch paper and its sequels by Golod. We have only considered here a small portion of the wonderful things they do.

It was mentioned in (N) that the question whether a right primitive ring need be left primitive was open. Recently G. Bergman has given an example of a ring which is right primitive but not left primitive. We now shall exhibit this example.

Let Q be the field of rational numbers and let $Q(x)$ be the field of rational functions in the indeterminate x over Q. In $Q(x)$ the mapping $\alpha: r(x) \to r(x^2)$ defines a monomorphism into but not onto. Let A be the ring of all polynomials $r_o(x) + r_1(x)y + \ldots + r_n(x)y^n$, with $r_i(x) \in Q(x)$ where equality and addition are as usual but where $yr(x) = r(x^2)y = \alpha(r(x))y$. A is a non-commutative integral domain where all left ideals are principal, since we have a left division algorithm. It is in A that we shall find the right, but not left, primitive ring R.

We first show

LEMMA 7.1. <u>Any subring</u> B <u>of</u> A, <u>containing</u> x <u>and</u> y, <u>which is an algebra over</u> Q <u>is right primitive.</u>

<u>Proof</u>. We shall exhibit a faithful, irreducible right B-module.

Given $r \in A(x)$ we define $r^* \in A(x)$ by

$$r^*(x^2) = \frac{r(x) + r(-x)}{2} .$$

We define a right A-module structure on $Q(x)$ by defining: $r \cdot y = r^*$, $r \cdot q = rq$ for $r, q \in Q(x)$. We claim this endows $Q(x)$ with the structure of a right A-module. We merely must check: for $r, q \in Q(x)$, $(r\cdot y)\cdot q = r\cdot(yq) = r(\alpha(q)y)$ $= (r\alpha(q))\cdot y$, that is, $r^*q = (r\alpha(q))^*$. Now $(r^*q)(x^2)$ $= r^*(x^2)q(x^2) = \frac{1}{2}(r(x) + r(-x))q(x^2) = \frac{1}{2}(r(x)q(x^2) + r(-x)q((-x)^2))$ $= \frac{1}{2}(r(x)\alpha(q)) + r(-x)\alpha(q(-x)) = (r\alpha(q))^*(x^2)$. Thus $r^*q = (r\alpha(q))^*$ and so we see that $Q(x)$ becomes a right A-module under the defined action.

From the definition of the action we immediately see:

$$\begin{aligned} x^n \cdot y^m &= x^{(n/2^m)} && \text{if } 2^m \mid n \\ &= 0 && \text{if } 2^m \nmid n \end{aligned}$$

Let M be the B-submodule of $Q(x)$ generated by x. We claim that M is both irreducible and faithful as a B-module.

Let $0 \neq \frac{p(x)}{q(x)}$ be in M where $p(x)$ and $q(x)$ are polynomials in Q. Let $n = \deg p(x)$; pick m so that $2^m > n$. Then $(\frac{p(x)}{q(x)})\cdot(q(x)x^{2^m-n}y^m) = (p(x)x^{2^m-n})y^m = cx$ where $c \neq 0$ is the leading coefficient of $p(x)$. Thus x is in the submodule generated by any non-zero element of M, in other words, since x generates M, M is irreducible.

Suppose now that $0 \neq b = \sum r_i(x)y^i$ in B annihilates M. Let $xp(x) \neq 0$ be a polynomial so that all $xp(x)r_i(x)$ are polynomials and let $0 \neq xp(x)b = \sum s_i(x)y^i$ where the $s_i(x)$ are polynomials. Thus $\sum s_i(x)y^i$ is in B and annihilates M. Let j be the least integer such that $s_j(x) \neq 0$ and let $d = \max(\deg s_i(x) - \deg s_j(x))$. Pick n so that $2^n \geq \deg s_j(x)$ and $2^{n-j^i-1} \geq d$. Let $s_j(x)$ have degree m. Consider $x^{2^n-m}\cdot(\sum s_i(x)y^i)$. In the term $x^{2^n-m}s_j(x)$ the highest power of x occuring with non-zero coefficient is x^{2^n} hence in $x^{2^n-m}s_j(x)\cdot y^j$, $x^{2^{n-j}}$ occurs with non-zero coefficient. For $i > j$ the highest power of x in $x^{2^n-m}s_i(x)$ is

$2^n - m + \deg s_i(x) \leq 2^n + d$ by definition of d, so in $x^{2^n-m}\, s_i(x)\cdot y^i$, x occurs at best to the power $\frac{2^n+d}{2^i}$; since

$$\frac{2^{n+d}}{2^i} \leq 2^{n-i} + d < 2^{n-j-1} + d \leq 2^{n-j-1} + 2^{n-j-1} = 2^{n-j},$$

in other words $x^{2^{n-j}}$ occurs with non-zero coefficient in $x^{2^n-m}(\sum s_i(x)y^i)$ whence $x^{2^n-m}\cdot(\sum s_i(x)y^i) \neq 0$, contrary to $\sum s_i(x)y^i$ annihilates M. This proves the lemma.

Let v_n be the valuation induced by the n-th cyclotomic polynomial $p_n(x)$ on $Q(x)$.

LEMMA 7.2. _If_ n _is odd_ $v_n(r(x)) = v_n(r(x^2))$.

Proof. Since $p_n(x)$ is irreducible over Q, any polynomial in $Q(x)$ having a common root with $p_n(x)$ has $p_n(x)$ as a factor. Now $p_n(x^2)$ for n odd has as roots $\pm\sqrt{\theta}$ where θ are the roots of $p_n(x)$ hence by the oddness of n, $p_n(x) | p_n(x^2)$. Similarly $p_n(-x) | p_n(x^2)$. Thus on comparing degrees, $p_n(x^2) = p_n(x)p_n(-x)$, hence $v_n(p_n(x^2)) = v_n(p_n(x))$. If $(p_n(x), q(x)) = 1$ then $(p_n(x), q(x^2)) = 1$ hence $v_n(q(x)) = 0$. But then for any polynomial $f(x), v_n(f(x)) = v_n(f(x^2))$. From this we get the result for rational functions and the lemma is established.

We define V_n on A, for odd n, by

$$V_n(\sum r_i(x)y^i) = \min v_n(r_i(x)).$$

From Lemma 7.2, $v_n(r(x)) = v_n(\alpha(r(x)))$ hence V_n defines a valuation on A. Let $R = \{z \in A | V_n(z) \geq 0 \text{ all odd } n\}$. Clearly R is an algebra over Q and contains x and y, hence by Lemma 7.1 R is right primitive. We now prove

THEOREM 7.5. R _is right primitive but not left primitive._

Proof. Let I be a maximal left ideal of R; we shall show that I contains a non-zero two-sided ideal of R, therefore R can not be left primitive.

Consider the left ideal AI of A; if $AI \neq A$ it is generated by some element $g = a_o(x) + a_1(x)y + \ldots + a_m(x)y^m$, $m > 0$, $a_m(x) \neq 0$, since the non-zero elements in Q(x) are invertible in A. Choose n odd so that $V_n(g) = 0$. Given any $a = \sum b_i y^i$ there is an i so that $V_n(a) = V_n(b_i)$; let $\delta(a)$ = largest i so that $V_n(a) = V_n(b_i)$. Clearly $\delta(ac) = \delta(a) + \delta(c)$ for all $a, c \in A$.

Now since $V_n(q) = 0$, $\delta(q) = m$ where m is the degree of g in y. If $u \in AI$ then $\delta(u) \geq \delta(g) > 0$. Let $p_n(x)$ be the n-th cyclotomic polynomial; $p_n(x)$ is in R and $V_n(p_n(x)) = 1$. Also $p_n(x) \notin I$ for it is invertible in Q(x), and so in A, and $AI \neq A$. Since I is maximal, $R = I + Rp_n(x)$ so for some $i \in I$, $r \in R$, $1 = i + rp_n(x)$; thus $1 - rp_n(x) \in I$. But $\delta(1 - rp_n(x)) = 0$, thus $0 = \delta(i) > 0$ since $i \in I$.

Therefore if I is a maximal left ideal of R then AI = A. Thus $1 = a_1 i_1 + \ldots + a_s i_s$ for $a_j \in I$, $i_j \in I$. We can find $w \neq 0 \in Q(x)$ so that for all n, $V_n(wa_i) \geq 0$ for $1 \leq i \leq s$. Thus $wa_i \in R$ hence $w = w(a_1 i_1 + \ldots a_s i_s) = (wa_1)i_1 + \ldots + (wa_s)i_s$ is an element of $RI \subset I$. Since $w \in I$, $Rw \subset I$. However, since w is invertible in A, $V_n(w^{-1}tw) = V_n(t)$ for all $t \in A$, in particular $w^{-1}Rw = R$ results. Thus $Rw = wR$ leading to $wR = Rw \subset I$. I thus contains the two-sided ideal wR of R. This proves the theorem.

We conclude this chapter with an example, due to Sasiada, of a simple radical ring.

Let F be any field and let A be ring of formal power series over F in the two non-commuting variables x and y. Let A' be all the elements in A whose constant term is 0.

It is easy to verify that A' is a radical ring. Suppose for the moment that we know that x is not in the two-sided ideal of A generated by $x - yx^2y$. Let U be an ideal of A' which contains $x - yx^2y$ and is maximal with respect to the exclusion of x. Let $R = A'/U$; R is a radical ring having a unique minimal ideal $S \neq (0)$ which is contained in every non-zero ideal of R and which contains $\overline{x} = x + U$. From this it is trivial that S is a radical ring devoid of non-trivial two-sided ideals. To know that S is simple we need but know that $S^2 \neq (0)$. But in R, $\overline{x} = \overline{y}\overline{x}^2\overline{y}$ where $\overline{y} = y + U$, so that $\overline{x} = (\overline{y}\overline{x})(\overline{x}\overline{y}) \neq 0 \in S^2$, hence $S^2 \neq (0)$. Thus S would be a simple radical ring. To produce the desired example we must merely show that x is not in the two-sided ideal of A generated by $x - yx^2y$. The proof we give here for this fact is from the recent paper by Cohn and Sasiada.

We introduce some notation. With A and A' as above, let M be the two-sided ideal of A generated by $u = x - yx^2y$ and let E be the set of elements in A having constant term 1. We begin with

LEMMA 7.3. *If* $c = \sum_1^m a_i u b_i$ *and* $b_1 \in A'$ *then there exist* $a_1' \in A$, $b_1' \in E$, $b_i' \in A'$ *for* $i > 1$ *such that*

$$c = a_1' u b_1' + \sum_2^n a_i u b_i' .$$

Proof. Since $b_1 \notin A'$, $b_1 = \beta b_1'$ where $\beta \neq 0 \in F$ and $b_1' \in E$; let β_i be the constant term of b_i. Then

$$a_1 u b_1 + \sum_{i=2}^n a_i u b_i = (\beta a_1 + \sum_2^n \beta_i a_i) u b_1' + \sum_2^n a_i u (b_i - \beta_i b_1').$$

This is the lemma with $a_1' = \beta a_1 + \sum_2^n \beta_i a_i$ and $b_i' = b_i - \beta_i b_1' \in A'$, for $i > 1$.

LEMMA 7.4. <u>If</u> $cx + dy = \sum_1^n a_i ub_i$ <u>with</u> $b_1 \notin A'$ <u>then</u> <u>there exist</u> $b_i^* \in A'$ <u>such that</u> $cx + dy = cu(1 + b_1^*) + \sum_2^n a_i ub_i^*$.

<u>Proof.</u> Using Lemma 7.3 we may suppose, given the hypothesis $b_1' \notin A'$, that $b_1 \in E$ and $b_i \in A'$ for $i > 1$, without changing the original a_i for $i > 1$. Write $b_1 = 1 + b_1'x + b_1''y$, $b_i = b_i'x + b_i''y$ for $i > 1$; substituting these in $cx + dy = \sum_1^n a_i ub_i$, using $u = x - yx^2 y$ and comparing coefficients of x yield $c = a_1 + a_1 ub_1' + \sum_2^n a_i ub_i'$. Now $1 + ub_1'$ is invertible in A and its inverse, $1 - ub_1' + (ub_1')^2 - \ldots + (-1)^n (ub')^n + \ldots$ is of the form $1 + ug$ for $g \in A$. Thus we have $a_1 = c(1 + ug) - \sum_2^n a_i ub_i'(1 + ug)$; substituting this value for a_1 in the given expression for $cx + dy$ yields

$$cx + dy = c(1 + ug)ub_1 - \sum_2^n a_i ub_i'(1 + ug)ub_1' + \sum_{i=2}^n a_i ub_i$$

$$= cu(1 + gu)b_1' + \sum_2^n a_i ub_i^* ,$$

where $b_i^* = b_i'(1 + ug)ub_1 + b_i$ is in A'. Since both b_1 and $1 + gu$ are in E, $(1 + gu)b_1$ is in E so of the form $1 + b_1^*$, $b_1^* \in A'$. This is precisely the assertion of the lemma.

LEMMA 7.5. <u>Suppose the product</u> dy <u>has the form</u> $dy = \sum_1^n a_i ub_i$, <u>where</u> n <u>is the minimal such for</u> dy. <u>Then each</u> b_i <u>is in</u> A'.

<u>Proof.</u> Suppose some $b_i \notin A'$, say b_1; using the previous lemma, with $c = 0$, yields $dy = \sum_2^n a_i ub_i^*$ contradicting the minimal nature of n.

We now come to the heart of the discussion. This is

LEMMA 7.6. M, <u>the ideal generated by</u> $u = x - yx^2y$, <u>contains no monomial</u>.

<u>Proof</u>. Suppose that the lemma is false; then there is some monomial $c \in M$, $c = \sum_1^n a_i u b_i$. For all such monomials pick a monomial c where n is minimal possible. For this n pick c of lowest degree lying in M. We assert that c must finish in x; if not, $c = c_1 y$. By Lemma 7.5 all the $b_i \in A'$ and writing $b_i = b_i'x + b_i''y$ and substituting in the relation for c gives $c_1 = \sum_1^n a_i u b_i''$ on comparing the coefficient of y. Since c_1 is of lower degree than c this is not possible. Hence we know that $c = c_1 x$ and $c_1 x = \sum_1^n a_i u b_i$.

If all the b_i are in A' writing $b_i = b_i'x + b_i''y$ will lead to the contradiction $c_1 = \sum_1^n a_i u b_i'$. Thus we may suppose $b_1 \notin A'$; by Lemma 7.4, $c_1 x = c_1 u d_1 + \sum_{i=2}^n a_i u d_i$ where $d_1 \notin A'$, $d_i \in A'$ for $i > 1$.

We claim that for any $r \leq n$ we can find monomials $p_1, \ldots, p_r$ of positive degree and elements $d_1, \ldots, d_n \in A$ (depending on r) such that

(1) $c_1 p_1 \cdots p_r x = c_1 \sum_1^r p_1 \cdots p_{i-1} u d_i + \sum_{r+1}^n a_j u d_j$,

(2) $d_i \notin A'$ for some $1 \leq i \leq n$.

Note that our preceding discussion established this for $r = 0$. We now proceed by induction.

Suppose for $r < n$ elements $d_1, \ldots, d_n$ and monomials $p_1, \ldots, p_r$ of positive degree can be found satisfying (1) and (2). We want to pass to $r+1$. We divide into two cases.

I. If $d_j \in A'$ for all $j > r$ then $d_{i_o} \notin A'$ for some $i_o \leq r$. Write $d_j = d_j'x + d_j''y$ for $j > r$ and $d_i = \lambda_i + d_i'x + d_i''y$ for $i \leq r$. Substituting in (1) and equating multipliers of x yields that

$$c_1p_1\cdots p_r = c_1\sum_1^r \lambda_i p_1\cdots p_{i-1} + c_1\sum_1^r p_1\cdots p_{i-1}ud_i' + \sum_{r+1}^n a_j ud_j' ;$$

hence if s is the least i such that $\lambda_i \neq 0$, on transposing we get

$$-\lambda_s p_1\cdots p_{s-1}(1+f) = \sum_1^r c_1p_1\cdots p_{i-1}ud_i' + \sum_{r+1}^n a_j ud_j' ,$$

where $f \in A'$ (since the p's are of positive degree). Transpose the s-th term of the first sum on the right to the left hand side . We then have

$$c_1p_1\cdots p_{s-1}(\lambda_s(1+f) - ud_s') = \sum_{\substack{i=1\\ i\neq s}}^r c_1p_1\cdots p_{i-1}ud_i' + \sum_{r+1}^n a_j ud_j' .$$

But $\lambda_s(1+f) - ud_s'$ is invertible in A giving us $c_1p_1\cdots p_{s-1}$ as a combination of $\sum a_i^* ub_i^*$ involving $n-1$ terms. This contradicts the choice of n and so case (I) is not possible.

II. Suppose, then that $d_j \notin A'$ for some $j > r$, say for $j = r+1$. Applying Lemma 7.4 to (1) gives

$$c_1p_1\cdots p_r x = \sum_1^r c_1p_1\cdots p_{i-1}ud_i + c_1p_1\cdots p_r(1+d_{r+1}) + \sum_{r+2}^n a_j ud_j ,$$

where all the d_k's, $1 \leq k \leq n$ are in A'. Writing $d_k = d_k'x + d_k''y$ and substituting in this leads, on comparing multipliers of x, to

$$(3)\quad c_1p_1\cdots p_r yx^2 = \sum_1^r c_1p_1\cdots p_{i-1}ud_i' + c_1p_1\cdots p_r ud_{r+1}' + \sum_{r+2}^n a_j ud_j' .$$

If in (3) all the d_k' are in A' we can write $d_k' = e_k'x + e_k''y$, substitute, pick out the multiplier of x to get

$$(4)\quad c_1p_1\cdots p_r yx = \sum_1^r c_1p_1\cdots p_{i-1}ue_i' + c_1p_1\cdots p_r ue_{r+1}' + \sum_{r+2}^n a_j ue_j' .$$

If in (4) all the e_k' are in A' we can repeat the by-now familiar argument to get

$$(5)\quad c_1p_1\cdots p_r y = \sum_1^r c_1p_1\cdots p_{i-1}uf_i + c_1p_1\cdots p_r uf_{r+1} + \sum_{r+2}^n a_j uf_j .$$

Since this is a minimal representation of $c_1p_1\cdots p_r y$, invoking Lemma 7.5, we have all the f's in A' and writing $f_i = g_i'x + g_i''y$ leads to a cancellation of y and the result

$$c_1p_1\cdots p_r = \sum_1^r c_1p_1\cdots p_{i-1}ug_i'' + c_1p_1\cdots p_r ug_{r+1}'' + \sum_{r+2}^n a_j ug_j'' ,$$

hence

$$c_1p_1\cdots p_r(1 - ug_{r+1}'') = \sum_1^r c_1p_1\cdots p_{i-1}ug_i'' + \sum_{r+2}^n a_j ug_j'' .$$

But $1 - ug_{r+1}''$ is invertible so we get a representation of the monomial $c_1p_1\cdots p_r$ in $n-1$ terms of the form $s_i ut_i$; this contradicts the choice of n. Hence in (3) or (4) some d_i' or some e_i' is not in A'. In either case this completes the induction.

The inductive step having been made we have that there exist $p_1, \ldots, p_n$ monomials of positive degree and $d_1, \ldots, d_n \in A$ such that

$$c_1 p_1 \cdots p_n x = \sum_1^n c_1 p_1 \cdots p_{i-1} u d_i \, .$$

But this puts us back into case I of our argument, which we saw was not possible. Hence the lemma is proved.

Since M contains no monomials it does not contain x. Hence, as we saw earlier, this forces the existence of a simple radical ring.

As Hamsher has pointed out this also enables you to construct a countable simple radical ring. Let F be countable; in A consider the subring generated by x and y -- it too is countable. Adjoin all the quasi-inverses of its elements to it and consider the subring now generated -- it is countable. Continue and take the direct limit to get a countable radical subring T of A containing x and y. In T, x is not in the ideal generated by $x - yx^2y$; now do as before, taking an ideal M of T maximal with respect to exclusion of x and containing $x - yx^2y$. The minimal ideal of T/M is a countable simple radical ring.

Bibliography

1. G. Bergman, "A ring primitive on the right but not on the left", Proc. Amer. Math. Soc., 15(1964), pp. 473-475.

2. P.M. Cohn and E. Sasiada, "An example of a simple radical ring, J. of Algebra, 5(1967), pp. 373-377.

3. E.S. Golod, "On nil algebras and finitely approximable p-groups", Izv. Akad. Nauk. SSSR Ser. Mat., 28(1964), pp. 273-276.

4. E.S. Golod and I. R. Shafarevitch, "On the class field tower," Izv. Akad. Nauk. SSSR Ser. Mat., 28(1964), pp. 261-272.

5. E. Sasiada, "Solution of the problem of the existence of a simple radical ring", Bull. Acad. Polon. Sci., 9(1961), pp. 257